Findings from Production Management Research

Series Editor

Peter Burggräf, International Production Engineering & Management, Universität Siegen, Kreuztal, Nordrhein-Westfalen, Germany

In the dynamic realm of industrial operations, production management remains the core for organizational efficiency. This series is a collection of doctoral thesis delving into the latest insights and innovations shaping production management, offering a comprehensive exploration of emerging trends and best practices. From the integration of advanced digital technologies to the effects of sustainability initiatives, this series covers key facets driving the evolution of modern production management. Drawing upon cutting-edge research and practical case studies, this collection is an indispensable resource for industry professionals, academics and students seeking to navigate and capitalize on the dynamic landscape of modern production management.

Ognjen Radišić-Aberger

Optimisation and Control of Engineering Change Schedules in the Automotive Industry with Metaheuristics and Machine Learning

Springer Vieweg

Ognjen Radišić-Aberger
Universität Siegen
Siegen, Germany

Dissertation zur Erlangung des Grades eines Doktors der Ingenieurswissenschaften vorgelegt von Dipl.-Ing. Ognjen Radišić-Aberger, B.Sc. eingereicht bei der Naturwissenschaftlich-Technischen Fakultät der Universität Siegen, Siegen 2025
Betreuer und erster Gutachter: Univ.-Prof. Dr.-Ing. Peter Burggräf, Universität Siegen
Zweiter Gutachter: Univ. Prof. Dr. phil. Ralph Dreher, Universität Siegen
Tag der mündlichen Prüfung: 12. November 2025

ISSN 3005-1649 ISSN 3005-1657 (electronic)
Findings from Production Management Research
ISBN 978-3-658-51073-2 ISBN 978-3-658-51074-9 (eBook)
https://doi.org/10.1007/978-3-658-51074-9

This Springer Vieweg imprint is published by the registered company Springer Fachmedien Wiesbaden GmbH, part of Springer Nature.
The registered company address is: Abraham-Lincoln-Str. 46, 65189 Wiesbaden, Germany

For Felix and Luzia

Acknowledgements This thesis on the applications of AI in Engineering Change Management was written during my time as a doctoral candidate within the PhD Programme at the BMW Group in collaboration with the University of Siegen, Germany. Within this timeframe, I received support from several colleagues and friends, whom I would like to thank.

Special thanks go to Univ. Prof. Dr.-Ing. Peter Burggräf, holder of the Chair for International Production Engineering and Management, for supporting this external thesis, providing me with the academic opportunity to realise this research. With intense talks and discussions, the challenging of my ideas, and provision of research views, the ideal input was provided to ensure a profound thesis.

I would also like to thank Univ. Prof. Dr. phil. Ralph Dreher, holder of the Chair for Technical Vocational Didactics, for his valuable input and willingness to serve as the second expert witness. Univ. Prof. Dr.-Ing. Tamara Reinicke, holder of the Chair for Product Development, for assuming the chair of the doctoral committee. Prof. Dr. rer. nat. Robert Brandt, holder of the Chair of Materials Systems for Lightweight Vehicle Construction for being a member of the doctoral committee.

I also thank my colleagues at the Institute, especially Fabian Steinberg, Alexander Becher, Till Saßmannshausen, and Johannes Wagner. Without our open discussions, the rigorous challenging of my work, and the creative talks, this thesis would not have come to be.

I am also grateful to my colleagues and managers at BMW, who gave me the time and resources for this research. Especially to my mentor, Karl Deutschmann. Without you planting the seed of thought of a potential doctoral thesis, I would have never started work on this chapter of my life. Greg Pirochta, Stephan Pichler, and Tobias Würzberger, without your support, your openness to new approaches, and input from years of industrial experience, this thesis would not have had the impact it did. A special thanks goes to Tim Weißer, my supervisor and friend, who ensured that my focus was always set right and ensured perfect research conditions in the applied field. I would like to thank the students I supervised during this time, Raquel, Dennis, and Maksym, whose own research helped when developing my own ideas. Finally, Mario and Christoph, thank you for buddying up. Our long talks and our shared experiences during the PhD programme ensured motivation through it all.

My deepest thanks go to my wife Elisabeth for her unwavering support during this time. Thank you, for providing me with the love and patience needed to start and finalise this thesis. Special thanks to my mother Vesna, my father Darko, and my sister Stella, whose life-long support gives me the freedom to pursue my goals.

Without all of you, this thesis would have never seen the day of light. Thank you.

Munich
November 2025

Ognjen Radišić-Aberger

Competing Interests The author has no competing interests to declare that are relevant to the content of this manuscript.

Abstract

Engineering changes (EC) are modifications to components, common to complex products, and require planning and thorough control of implementation schedules to reduce economic downsides. However, determining cost-optimal EC effectivity dates is difficult due to fluctuating demand and production figures. Furthermore, mismanaged EC processes frequently lead to off-schedule ECs, invoking costs of quality and stock obsolescence. Due to EC inherent complexity, however, capacity for thorough EC control is rarely sufficient and prioritisation by likelihood of failure could help focus on high-risk ECs. For both challenges, optimisation and control of EC schedules, recent advancements in artificial intelligence (AI) could provide enhanced computational support. Nevertheless, analysing existing AI-artefacts for EC management as part of a systematic literature review showed two main research gaps: Firstly, existing models are incapable of simultaneous optimisation of effectivity dates for associated ECs, necessitating applied research of metaheuristics; secondly, EC schedule control could be improved by research and application of new machine learning (ML) algorithms. Therefore, as a main outcome of this thesis, it is shown how AI can be applied for effective and efficient optimisation and control of EC schedules. In detail, specific models for EC effectivity date optimisation and EC effectivity date adherence prediction were developed and subsequently validated at an industry partner in a case-based approach. Firstly, a new mathematical model for optimisation of EC effectivity dates was developed and tested with five common metaheuristics. Based on the specific results, it was shown how a combination of different algorithms can efficiently calculate, monitor, and refine optimal effectivity dates even for extensive ECs. Subsequently, to determine the chances of an EC being off-schedule, methods of ML and predictive business process monitoring were applied on predicting adherence to EC effectivity dates. Testing different ML approaches, it was indicated that given sufficient data engineering, ML models can accurately forecast the adherence to planned effectivity dates and enable early countermeasures. Following up on these initial findings, it was further shown that ML models can predict adherence to EC schedules in the earliest EC process phases. Moreover, it was presented how responsiveness to data generated during process execution influences predictive performance and thus its inclusion into the data model has been recommended. In conclusion, the prediction and optimisation models developed within this thesis provide reasonable support to change coordinators and material planners with scheduling and controlling EC effectivity dates, reducing capacity constraints for EC within organisations.

Zusammenfassung

Technische Änderungen (TÄ) sind Modifikationen bestehender Produkte, deren Einsatzterminierung insbesondere in der Produktion komplexer Produkte sorgfältige Planung und Steuerung zur Vermeidung wirtschaftlicher Nachteile erfordert. Produktions-schwankungen erschweren jedoch die Ermittlung eines kostenoptimalen Einsatztermins. In weiterer Folge führen ungeregelte TÄ-Prozesse zu Einsatzterminverfehlungen welche erhöhte Qualitäts- und Verschrottungskosten hervorrufen. Die verfügbare administrative Kapazität für eine durchgehende Überwachung des TÄ-Prozesses ist jedoch aufgrund der dem TÄ-Management inhärenten Komplexität selten ausreichend, weswegen Priorisierungsmethoden zur Fokussierung genutzt werden. Mit den fortschreitenden Entwicklungen in der Künstlichen Intelligenz (KI), ergeben sich allerdings neue Möglichkeiten zur Optimierung und Überwachung der Einsatzterminierung. Eine Analyse von KI-basierter TÄ-Literatur zeigte zwei primäre Forschungslücken auf: Einerseits erfordert die simultane Optimierung von Einsatzterminen mehrerer zusammenhängender TÄ anwendungsorientierte Forschung im Bereich der Metaheuristiken. Andererseits ermöglicht die Erforschung und Anwendung von neuen Methoden des Maschinellen Lernens (ML) eine verbesserte Prozessüberwachung. Dahingehend wird in dieser Arbeit die Anwendbarkeit von KI-Methoden für die Optimierung und Überwachung des TÄ-Einsatztermins in der Produktion erforscht. Im Detail wurden in einem fallbasierten Ansatz Modelle zur Optimierung des Einsatztermins sowie Vorhersagemodelle zur Überwachung der Einsatztermintreue entwickelt und bei einem Industriepartner validiert. Zur Termin- und Bestandsoptimierung wurden fünf gängige Metaheuristiken mit einem neuentwickelten mathematischen Modell getestet. Basierend auf den fallspezifischen Ergebnissen wurde gezeigt, dass eine Kombination unterschiedlicher Algorithmen selbst für umfangreiche TÄ effizient optimale Einsatztermine berechnen und nachjustieren kann. Zur Überwachung und Sicherstellung des Einsatztermins wurden Methoden des ML und des Predictive Business Process Monitoring zur wahrscheinlichkeitswertbasierten Vorhersage verwendet. Durch die Erprobung verschiedener ML-Ansätze wurde gezeigt, dass bei ausreichender Datenaufbereitung ML-Modelle die Termintreue hinreichend genau vorhersagen können. Dies ermöglicht eine frühzeitige Maßnahmendefinition zur Kostenreduktion. In Ergänzung zu diesen ersten Erkenntnissen wurde weiters dargelegt, dass ML-Modelle die Termintreue bereits in frühesten Prozessphasen vorhersagen können. Darüber hinaus zeigte sich, dass die Sensitivität auf dynamische Prozessdaten die Vorhersageleistung beeinflusst, weswegen deren Aufnahme im Datenmodell empfohlen wird. Zusammenfassend lässt sich festhalten, dass die in dieser Dissertation erarbeiteten Optimierungs- und Vorhersagemodelle ein deutliches Potential zur Reduktion des administrativen Aufwands in der Material- und Änderungssteuerung zeigen.

Contents

Abbreviations

AAGG	Advanced Aggregation Based Encoding
ACO	Ant Colony Optimisation
AGG	Aggregation Based Encoding
AI	Artificial Intelligence
AUC	Area under the Curve
AUC-PR	Area under the Precision Recall Curve
AUC-ROC	Area under the Receiver Operator Characteristic Curve
BN	Bayesian Network
BNN	Branching Neural Net
BuNe	Business Need
CAGG	Complex Aggregation Encoding
CCB	Change Coordination Board
CL	Case Length
CPA	Communication, perceiving, and acting
CPM	Change Prediction Method
CRISP-DM	Cross Industry Standard Process for Data Mining
DL	Deep Learning
DSM	Design Structure Matrix
DSR	Design Science Research
DSRM	Design Science Research Methodology
EC	Engineering Change
ECM	Engineering Change Management
ETO	Engineer to Order
GA	Genetic Algorithm
g_i	feature vector
I(n)	Insight (n)

IDX	Index Based Encoding
IS	Information System
KPI	Key Performance Indicator
KRP	knowledge, reasoning, and planning
LSTM	Long-Short-Term-Memory
MAS	Multi Agent System
MCC	Matthew's correlation coefficient
MH	Metaheuristic Search Algorithm
ML	Machine Learning
MP	Materials Planner
NLP	Natural Language Processing
NN	Neural Net
NPD	New Product Development
O(n)	Observation (n)
OEM	Original Equipment Manufacturer
PBPM	Predictive Business Process Monitoring
PoC	Proof of Concept
PR	Precision Recall Curve
PS	Problem-solving
PSO	Particle Swarm Optimiser
QC	Quality Criteria
R2	Coefficient of Determination
REQ	Requirement
RF	Random Forest Algorithm
RF_aagg	ML model trained with Random Forest algorithm on advanced aggregation encoded data
RF_agg	ML model trained with Random Forest algorithm on aggregation encoded data
RF_cagg	ML model trained with Random Forest algorithm on complex aggregation encoded data
RF_idx	ML model trained with Random Forest algorithm on index encoded data
ROC	Receiver Operator Characteristic
RQ	Research Question
S(n)	Search string (n)
SA	Simulated Annealing
SOM	Self Organising Map
TS	Tabu Search
UKR	Uncertain knowledge and reasoning

XGB	XGBoost Algorithm
XGB_aagg	ML model trained with XGBoost algorithm on advanced aggregation encoded data
XGB_agg	ML model trained with XGBoost algorithm on aggregation encoded data
XGB_cagg	ML model trained with XGBoost algorithm on complex aggregation encoded data
XGB_idx	ML model trained with XGBoost algorithm on index encoded data

Symbols

Symbol	Unit	Description
AUC-ROC	–	Area under the Receiver Operator Characteristic Curve
CD	–	Critical Distance
cfn	€/$	Cost of false negatives
cfp	€/$	Cost of false positives
CostBatch	€/$	Total Cost for EC batch
CostEC	€/$	Total Cost for EC
D	–	set of possible dates for the effectivity date
d0	TU	earliest possible call-off date
de	TU	EC effectivity date
DL	TU	Delivery lead time
dl	TU	Phase out date
F	–	set of factories affected by an EC batch
FNR	–	False Negative Rate
FPR	–	False Positive Rate
IC	€/$	Total Sum of inventory costs
J	–	set of components affected by an EC batch
NEC	€/$	Normalised Expected Cost
p(+)	–	proportion of positive examples
PC	€/$	Total Sum of penalty costs

(continued)

(continued)

Symbol	Unit	Description
PC	–	Probability Cost
RC	€/$	Total Sum of running costs
s0	–	Current stock
t	–	decision threshold
TPR	–	True Positive Rate
TS	–	Temporal Stability
X_F^2	–	Friedman statistic
Xdel	–	Volume of prechange component delivered until the effecitivty date
Xpre	–	Volume consumed of prechange component
yl	–	Delivery quantity
$\hat{y}$	–	Predicition score
z	–	skew

List of Figures

List of Tables

1 Introduction

To maintain competitiveness, a company's products must continuously align with evolving market demands and customer preferences. Frequently, these requirements cannot be met without modifying the products. As product life cycles shorten, software usage increases, and supply chains become more volatile, the frequency of necessary changes intensifies [2]. Due to product diversification, the frequency of changes further combines with an ever-increasing complexity in products and processes [3, p. 19]. This surge in interfaces and interdependencies makes managing changes progressively more challenging. These adaptations and changes are known as engineering changes (EC) [4]. Effectively optimizing and controlling these changes is critical for a company's long-term success [5]. Failure to address the challenges of engineering change management (ECM) can lead to severe economic and legal consequences. Up to 30% of a product's price and a 20% increase in warranty costs can be attributed to ineffective EC control [6, 7]. In extreme cases, such as the General Motors recall, a poorly documented change resulted in the recall of 30 million vehicles and multiple fatalities [8, 9]. In general, ECs are a major cause of disruptions in engineering and production processes. This is especially true in high-volume complex product manufacturing industries such as the automotive sector [10, 11]. To prevent issues and to reduce the impact of ECs, research in the ECM domain has focused on five goals: *less, earlier, more efficient, more effective, and better* [12].

O. Radišić-Aberger, *Optimisation and Control of Engineering Change Schedules in the Automotive Industry with Metaheuristics and Machine Learning*, Findings from Production Management Research,
https://doi.org/10.1007/978-3-658-51074-9_1

1.1 Challenges in Engineering Change Management

Managing ECs, however, presents itself with two major challenges: high complexity in ECM and limited administrative capacity. The complexity is primarily driven by the phenomenon of change propagation [13]. This phenomenon refers to the spread of ECs through the entire product along interfaces. Here, a change in one component may require a corresponding change in a related one. As an example, an EC that increases the diameter of a car's exhaust pipe due to thermal constraints could necessitate a change to the rear bumper to ensure collision-free assembly. The modified rear bumper could then need new mounting points, requiring ECs in the body work, and so on. Due to change propagation, predicting the full impact of even a small change on the product and production system is near impossible [14]. s a result, all affected ECs must be implemented simultaneously to preserve the fit and function of all impacted components. In mass-production industries, such as the automotive industry, change propagation can significantly disrupt ongoing production operations. To ease the management of ECs, the principal strategy of implementing ECs grouped into batches has been devised [15]. However, even with this strategy, the required administrative effort for managing ECs is overburdening. From the initial EC creation phase, the subsequent solution and approval phases, and ultimately the EC implementation phase, ECs demand dedicated capacity. Studies report of individual companies processing more than 100 000 ECs per year [11] and ECs causing up to 30% of an engineer's workload [12, 16]. Likewise, as shown by WÄNSTRÖM AND JONSSON [17], to reduce EC induced logistics costs material planners control delivery dates and lot sizes on a daily base. Thus, automation and computational EC support yield high benefits for any industry.

Computational support in ECM, however, has always been limited by the complexity inherent to ECs [18]. By Ashby's law, any controlling system would require at least the same level of complexity as the controlled system [19]. Translated, any system supporting EC requires at least the same level of complexity as ECs. Existing IT-tools, however, often fail to sufficiently represent the actual complexity of ECs [10, p. 99]. Nevertheless, recent ECM research has shown how artificial intelligence (AI) can provide a potential path to mitigate challenges and increase the level of automation [16, 20].

Within the field of ECM, several AI-applications have already been researched and computational support in ECM improved significantly [20, 21]. However, the EC implementation phase remains a white spot [22]. Especially two tasks hold a high potential for improvement by AI: optimisation and control of the EC implementation schedule. Regarding the first task, optimisation of the EC implementation schedule, the objective is to reduce EC implementation related costs by defining an optimal EC implementation schedule [4]. This is done by managing delivery dates and delivery quantities in relation to actual production figures, such that the total cost of stock obsolescence and quality is at a minimum on the resulting EC effectivity date. This effectivity date is easily identified for ECs affecting only one component. For associated ECs within an EC batch, however, the EC implementation optimisation is complex, as multiple factors must be considered simultaneously. Moreover, delivery dates, quantities, and production figures are only available as discrete aperiodic functions – adding complexity to the mathematical problem. Currently employed methods optimise the EC effectivity date by means of linear programming optimisation and linear approximations of stock and production figures [23, 24]. This, however, has two main downsides: Due to method inherent restrictions it is impossible to calculate the EC effectivity date for entire EC batches, and incorporating the actual delivery, stock, and production figures [25, p. 530]. Moreover, as stock and delivery quantities are described by discrete aperiodic functions, even advanced integer programming methods cannot be applied. Thus, for the first task, there is a need for an EC effectivity date optimisation model, which can efficiently calculate with discrete aperiodic input functions. For the second task, control of EC implementation according to schedule, the challenge is similar. Although an optimal EC effectivity date has been identified, unplanned process deviations can lead to an off-schedule EC. As capacity is limited, EC research suggests to focus on ECs with specific indicators, such as high cost [26], high impact [27] or high complexity [15]. However, all these approaches have the downside that ECs which fall into a lower category could be overlooked, or unnecessary effort put into ECs which would be implemented on time anyway. Other domains have resolved this by data-based decision making based on predicted outcomes or values [28]. Transferred, it is suggested to analyse whether the adherence to planned effectivity dates of an EC can be predicted. Within process literature, it is consensus that to predict the outcome of a process, a thorough understanding of the process and its current state is required [29]. However, as the EC process is rarely executed according to a formalised process [30, 31], it is difficult to define a rule set for interventions or predefine all possible future process sequences [32]. Furthermore, the data required to draw information from is spread across multiple departments [30, 33] and outcome defining process steps are difficult to pinpoint. For optimisation and control tasks with these characteristics, the research field of AI provides two groups of algorithms: Metaheuristic search algorithms (MH) for efficient optimisation of mathematically complex problems and machine learning (ML) for control by predicting future outcomes from past input-outcome pairs without previous knowledge of dependencies [34]. Nevertheless, performance of these algorithms has been shown to be case sensitive [34–36], necessitating domain specific development and validation of optimisation and prediction models. Overall, regarding the control of EC implementation according to schedule, it can be stated that there is an unfulfilled demand for advanced optimisation and prediction models to thoroughly optimise and control EC implementation.

In conclusion, there is a clear research gap in evaluating the applicability of AI-methods for optimisation and control in the EC implementation phase. With the ECM goals of introducing ECs earlier, more efficient, and more effective, computational support is necessary to alleviate capacity constraints. However, present methods have shown to be unsuited to manage domain inherent complexity. Here, emergent AI methods pose a new possibility. To evaluate whether these methods are suitable for optimisation and control of EC schedules, additional research is required. Thus, the main research question of this thesis can be formulated: *Can AI efficiently and effectively support in optimising EC effectivity dates and controlling adherence to these?* To provide an answer, this dissertation presents the results of applied research aimed at providing blueprints for application of AI methods in ECM, targeted at the EC implementation phase. Firstly, the performance of different MHs for efficient optimisation of EC effectivity dates and inventory will be examined. Secondly, ML-model performance for predicting the adherence to EC effectivity dates and thus to control the EC implementation according to schedule will be evaluated. Lastly, by applying predictive business process monitoring (PBPM) methods, it will be shown how ML can predict the adherence to EC effectivity dates from initial planning until final implementation.

The remainder of the thesis is structured as follows: Chapter 1 introduces the problem of EC implementation, the research methodology and the research objectives. Chapter 2 reviews existing EC implementation literature and provides definitions within the ECM domain. As this cumulative thesis is formed by four publications (Chaps. 3–6), to reduce redundancy, an overview of AI is given in Chap. 3.3.2. The basics of the algorithms and methods applied are likewise presented in the respective Chapters. The four publications are reproduced with standardised formatting, citation style, and spelling. Finally, a critical reflection and summary is provided in Chap. 7 and 8 respectively.

1.2 Research Methodology

ECM as a research field is driven by observations of real-world problems encountered during design and new product development. Integral studies examine existing and develop new methods to improve engineering design and production management by providing guidelines for human decision-making in organisations. Therefore, according to the definitions by CARNAP [37], ECM research can be attributed to the applied sciences, as opposed to the basic sciences. Consequently, this thesis also falls into the scientific domain of applied sciences. Adding to the traditional design and production science defining ECM research, with the added overlay of AI, computer science comes in as a third research area to intersect. Thus, any research methodology applied must cover all three research areas.

To encompass the design, production, and computer sciences, the primary research methodology of this work follows the design science research (DSR) methodology by Hevner et al. [38]. It has its roots in engineering and the science of information systems (IS), providing a systematic method to create new scientifically meaningful and economically relevant design artefacts. At its core, as opposed to other design research methods, DSR requires to root each artefact in observed real-world problems. These real-world problems can then be used to derive distinct business needs (BuNe), which are fulfilled by the developed artefact [38, 39]. Three main building blocks define the DSR Methodology: *environment*, *knowledge base* and at their intersection *IS research,* which in turn consists of developing and justifying the resulting artefact (Fig. 1.1).

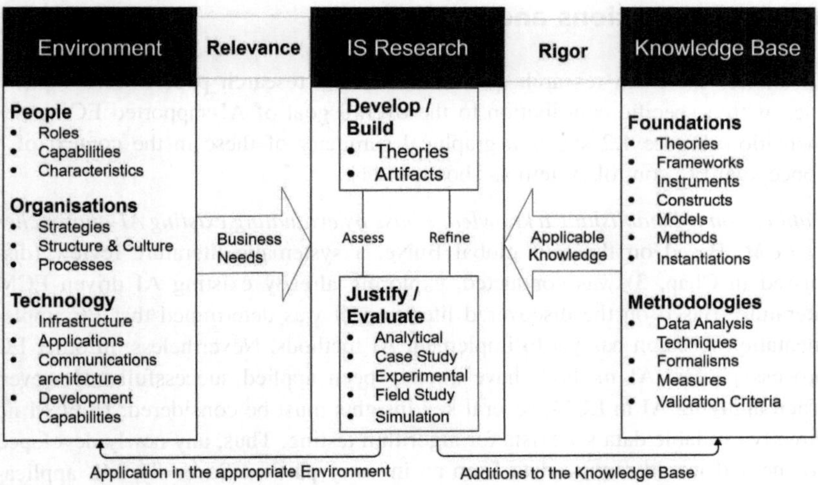

Fig. 1.1 The applied design science research methodology according to Hevner et al. [38]

Defining the environment for which the artefact is designed, is the root for any relevancy of any developed solution, represented by BuNe. As the observed organisation, technology, and people vary on case base, so does the exact definition of the resulting BuNe. However, as shown in Chap. 2, four global BuNe in EC implementation can be determined: *optimisation*, *decision support*, *supervision*, and *automation*. Within this thesis, the first three are addressed.

To add scientific rigor, the knowledge base plays an essential role. Already existing artefacts provide the foundations for any new IS research. The principal goal is to understand already existing solutions and to identify possible transferable artefacts. Methodologies on the other hand provide rigor during evaluation, to appropriately assess the newly developed artefacts. Within this thesis, a general knowledge base on ECM and EC implementation is provided in Chap. 2 with an added layer of AI in Chap. 3. For the development of artefacts in Chaps. 4, 5, and 6, a case relevant knowledge base is added.

Finally, at the core of DSR is the design of new artefacts. Here, innovative artefacts are developed to address the BuNe identified in the environment, enhanced with available knowledge, and improved through iterative circles. As each artefact addresses a specific BuNe, their design process is also individually discussed in each respective chapter.

1.3 Publications and Outline

To answer the main research question, four full research papers were written, each with a specific contribution to the overall goal of AI-supported EC implementation. Figure 1.2 shows a graphical summary of these in the context of a conceptual EC control system as shown in [1].

Publication I: Establishing a knowledge base by evaluating existing AI-approaches in ECM. Based on the four global BuNe, a systematic literature review (discussed in Chap. 3) was conducted, exploring already existing AI driven ECM literature. Based on the discovered literature, it was determined that EC implementation research has yet to implement AI methods. Nevertheless, in other EC process phases, AI methods have already been applied successfully. However, when applying AI in ECM, several key insights must be considered: In ECM no directly available data set exists for algorithm testing. Thus, any newly developed AI method would require data from an industry partner. Secondly, ML applications behave differently in ECM. The required data for successful prediction of

EC impacts is comparably higher to those applications in other domains. Furthermore, EC data has shown to be often available only in textual form, showing the necessity to employ data mining and process mining techniques for extraction of data.

Publication II: Benchmarking metaheuristics for EC effectivity date determination with a newly developed optimisation model. The second publication relates to the BuNe of optimisation. The date on which a change is implemented, known as the EC effectivity date, has previously been calculated with linear programming methods or brute force algorithms. However, considering that EC are mostly introduced in batches or releases, the previously simple problem becomes a numerically complex problem. Optimising the incoming and outgoing goods quantities, represented as inverted saw curves, becomes computationally expensive if an exact solution is required. To advance, a metaheuristics approach was developed and tested with EC data from an automotive manufacturer. Based on a newly defined mathematical model which incorporates different EC related costs, five common algorithms were tested (Genetic Algorithm (GA), Ant Colony Optimisation (ACO), Particle Swarm Optimisation (PSO), Tabu Search (TS), and Simulated Annealing (SA)). Of the tested algorithms, GA showed the best performance regarding solution quality and reliability. However, due to its comparatively long calculation time, it was suggested to combine the GA for the initial determination of an effectivity date and use TS for monitoring the costs. The developed artifact thus shows how metaheuristics can indeed efficiently optimise the EC effectivity date. Furthermore, the developed mathematical model allows for quick addition of other EC related costs, enabling their further reduction.

Publication III: Testing shallow and deep learning algorithms' predictive performance on ECM data for EC schedule adherence prediction. Once an EC effectivity date is determined, it should be checked whether it is achievable. Often, the validity of these dates is determined based on personal experience from change coordinators and material planners. However, once a date has been defined, its plausibility is not regularly rechecked. This leads to a situation where required tasks are delayed or even forgotten, and expensive emergency counter measures must be taken. To avoid this, publication III investigates whether the adherence to a chosen effectivity date can be determined by ML algorithms. For this, four ML algorithms (Random Forest, XGBoost, a Long-Short-Term-Memory Neural Network and a Branching Neural Network) were applied on data from an industry partner. As this task is set in a business context, the evaluation method for predictive performance was also changed from common receiver-operator-characteristic

(ROC) curves to cost curves. The research has shown that the applied shallow learners (Random Forest and XGBoost) are capable of effectively predicting the adherence to planned EC effectivity dates. The resulting artifact, validated with the industry partner, further shows how the ML-models can help prioritise scarce administrative capacity to high-risk cases.

Publication IV: Evaluating machine learning models' early predictive performance for monitoring EC schedule adherence from start to end. Finally, publication IV discusses how early the adherence to effectivity dates can be predicted. Comparing XGBoost, Random Forest and Long-Short-Term-Memory Neural Net algorithms, it was discovered that an XGBoost approach predicts most accurately (measured as AUC-ROC) for the biggest bulk of EC cases. However, it is shown that the prediction score fluctuates comparably more compared to Random Forests due to beneficial responsiveness to dynamic process data. The research outcome thus shows that it is possible to predict the adherence to EC effectivity dates as early as the initialisation of the EC process. Integrating the ML model into an EC schedule monitoring tool at the industry partner further shows how effective and efficient EC process control can be achieved at early process phase through the application of ML.

To conclude, this thesis offers insights on how AI can support associates even in complex business processes by example of the EC process. By dividing the general task into manageable problems, three artefacts were developed to support human decision makers in managing EC implementation.

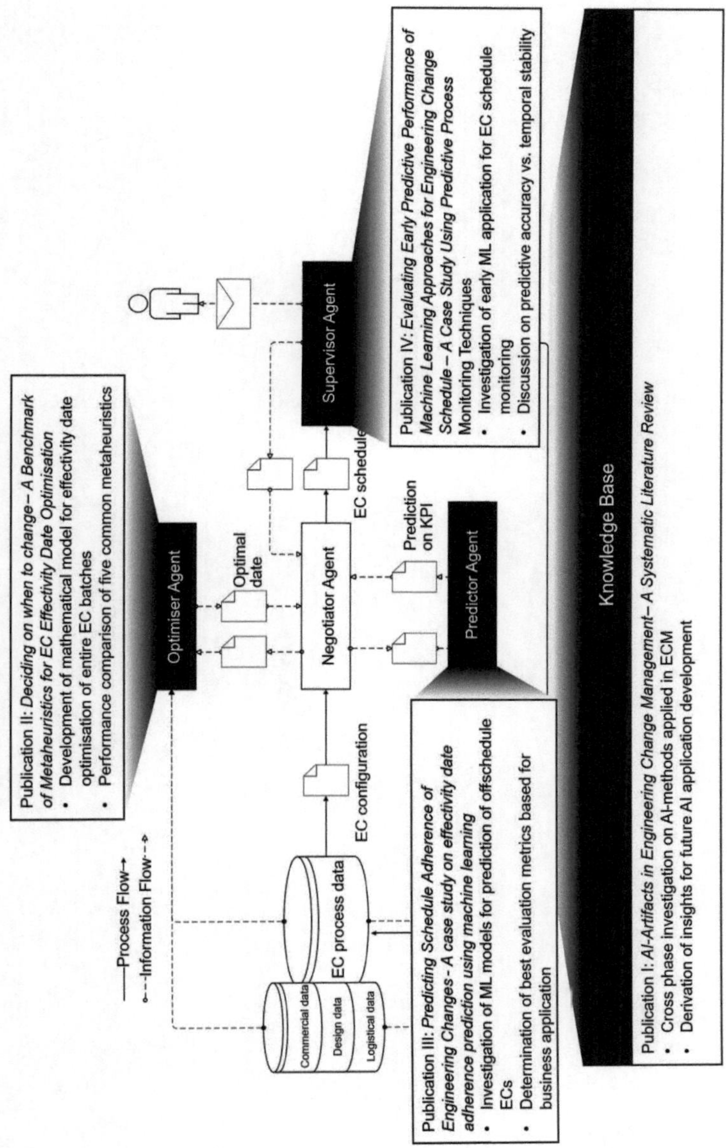

Fig. 1.2 Relation of the 4 main publications to the Multi Agent System in RADISIC-ABERGER ET AL. [1]

Theoretical Background 2

To provide a common base for discussion and deepen the understanding of challenges in ECM, basic definitions and theoretical background are required. Section 2.1 introduces definitions for EC, ECM and effectivity dates used throughout all publications. The accompanying EC process is further described in Sect. 2.2, whereas Sect. 2.3 displays current literature for EC implementation support.

2.1 Engineering Change

As presented, ECs and their proper control are essential to stay profitable. Especially in complex products, a design is rarely done from scratch but rather an adaptation of previous designs [40], leading to efficient EC management being a competitive advantage [12]. To remain consistent with previous literature, in this thesis the definition of ECs as given in [4, p. 475] is applied:

> *"ECs are changes and/or modifications to released structure (fits, forms and dimensions, surfaces, materials etc.), behaviour (stability, strength, corrosion etc.), function (speed, performance, efficiency, etc.), or the relations between functions and behaviour (design principles), or behaviour and structure (physical laws) of a technical artefact".*

By applying the definition as above, ECs in the context of this thesis are limited to modifications of a technical artefact e.g. a part or product of mechanical design. ECs on non-technical artefacts such as construction buildings are excluded due to a difference in the production system. Additionally, EC in the context of

O. Radišić-Aberger, *Optimisation and Control of Engineering Change Schedules in the Automotive Industry with Metaheuristics and Machine Learning*, Findings from Production Management Research,
https://doi.org/10.1007/978-3-658-51074-9_2

circuit design is excluded, as this research focuses on methods to enable late logic changes by usage of flexible design which is oriented more towards design principles and less towards ECM. These areas have developed their own research streams, and usage of AI is focused more towards requirements management [41] or design optimisation [42].

In this context, ECM describes how an organisation handles ECs [43]. The goals of ECM have been summarised and described as: less, earlier, more effective, more efficient, and better [12]. Less can be interpreted as frontloading, as ideally a change is totally avoided. However, when discussing the implementation phase, avoidance has already been omitted, and the EC must occur. Thus, the goals of earlier, more effective, and more efficient come into focus. In the context of the implementation phase, earlier does not mean necessarily that an EC must be implemented earlier, but that steps towards a beneficial introduction are undertaken as soon as possible. More efficient and more effective are two goals which lie at the core of EC implementation research [4]. They can be described as the intention to use less resources at less costs, all while having a sustainable impact on EC implementation. Finally, the goal better is a retrospective point of view, comparable to a feedback loop.

Most ECM research concludes that communication among stakeholders and supervising the process are the key to increased efficiency. Moreover, with the prevalence of module-based engineering [15] and distributed production networks in complex industries, the EC implementation phase receives an additional increase in complexity and attention, as a multitude of suppliers and production facilities have to be managed. As a result, the communication effort increases further, while capacity remains stagnant. Hence, it is suggested to investigate additional possibilities for computational support, possibly with the use of AI [44].

2.2 Engineering Change Process

Within literature, no commonly agreed or ideal EC process exists, mainly due to the specifics of the organisations implementing them [30]. However, the general EC process model by Jarratt [45] has established itself as a good overview process model. Here, the main phases of any EC can be seen: The start at its trigger, the design phase, its approval, followed by its realisation (Fig. 2.1).

Every EC is initiated by a trigger, in the scientific community described as emergent or imminent depending on the source of the trigger [14]. Following the initial change request, the EC is classified, a solution developed, and its impact assessed (steps 1–3). Early phase ECM research is driven by understanding the cause of EC, identifying change propagation paths, and reducing the required effort of design engineers. During this early phase, the term engineering change request or engineering change proposal [46] is used for the associated paperwork. It is a form open to any employee to raise problems or optimisation potentials

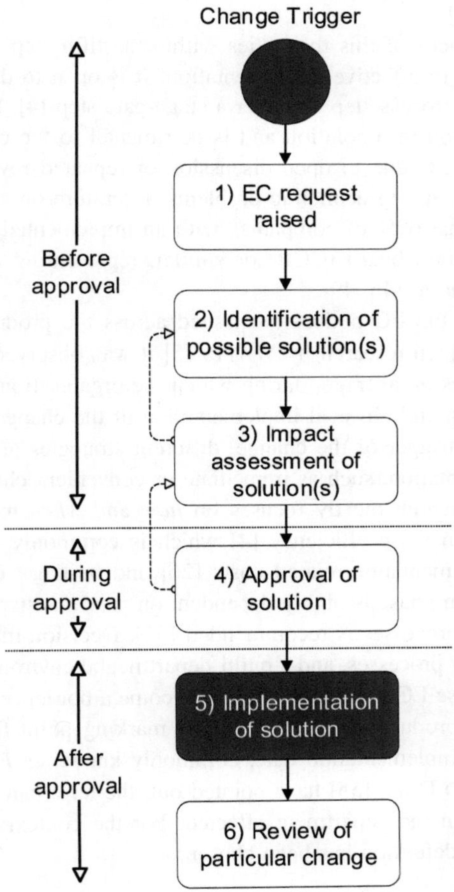

Fig. 2.1 General EC Process as described by JARRATT ET AL. [46]. The highlighted 5th step shows the focus of this thesis

for a given product [47]. After the fourth step, this request is transformed into an engineering change order or engineering change notice, initiating the directive to introduce the change to production [47]. The focus lies in implementing the change to manufacturing and assembly with as little disturbance to the production process as possible. Change implementation is followed by a documentation and review process step, essential for performing better in the future. Research in the last process phases aims at organising knowledge to provide feedback to optimise future changes [43].

Although the focus of this thesis lies within the fifth step, the approval step is a key milestone to effective implementation. It is open to discussion whether it is a stand-alone process step or rather a stage-gate step [4], however, approval leads to the selection of a solution and is detrimental to the outcome of an EC [48]. A held-back EC due to open discussion or repeated revision can delay a necessary change, causing economic problems. Literature on approval processes in EC concludes that 60% of companies with an implemented ECM system use a change coordination board (CCB) or similar, representing relevant functions within the organisation [46, 49].

After approval, the EC notices are issued across the production and supply chain network [48]. In research by WATTS [5] it was observed that the ensuing phase takes 40 days on average, during which the organisation performs various tasks for realisation and physical implementation of the change [50]. Depending on the nature and trigger of the change, different strategies are applied when it comes to implementation, such as immediate or convenient change [23, 51]. EC implementation research mostly focuses on *how* and *when* to implement, with the main goal to increase efficiency [4] which is commonly interpreted as the reduction of implementation-related costs [26] and best use of resources [52]. The implementation phase is highly dependent on previous steps, and integration into the upstream processes is recommended [53]. Decision-making with partial knowledge, lengthy processes, and a multi-departmental environment without visible ownership cause EC implementation to become a bottleneck process, instead of an enabler for product maturity [54]. The marking point for the end of the fifth phase is the implementation date, commonly known as *EC effectivity date.* As BALCERAK AND DALE [55] have pointed out, the term can relate to different dates, depending on the department affected. For the context of this thesis, an assembly-oriented definition has been chosen:

> *"The EC effectivity date is the date, after which no pre-change component and before which no post-change component is used for assembly".*

This fluid definition is especially important for the case studies of this thesis, as in mass-customisation industries, such as the automotive industry, components might not be used for production every day.

2.3 State of the Art in Engineering Change Implementation

As previously stated, the fifth phase, EC implementation, consist of all tasks to realise the change. Within this area, three research approaches emerged: case studies, optimisation strategies, and process modelling. In the following, relevant contributions and methods targeting this process phase are discussed, clustered by methodology.

Though initial contributions are allocated into different groups of the framework described in [4] (see Fig. 3.1), most EC implementation research builds on qualitative studies by DIPRIMA [56], REIDELBACH [57], and an analytical study by BALAKRISHNAN AND CHAKRAVARTY [51] which introduce basic strategies and profit calculations. Where [56] describes the process a change coordinator follows, the second contribution determines the factors affecting long lead times for change implementation [57]. Additionally, DIPRIMA [56] names some of the task which are to be done during EC implementation, amongst others the determination of gross requirements, initiating orders, and scrapping non-usable obsolete components.

BALAKRISHNAN AND CHAKRAVARTY [51] were one of the first researchers who introduced different implementation scenarios. These scenarios discuss the possibility of an EC having an *'arbitrary introduction date'*, a *'firm planned orders scenario'* (where changed and pre-change parts coexist), and finally *'a use-up technique'*, all based on functions within manufacturing resource planning systems. By providing variables with monetary values and a set of formulas, they produce an optimal cut-out-cut-in date, also known as effectivity date, targeting a maximised overall profit. Building on this contribution, several authors have provided expanded research in EC implementation and the three branches of research become more distinguishable.

2.3.1 Case Studies in Engineering Change Implementation

EC research not only requires validation with industry partners but also case studies to discover if developed methods are still relevant [58]. With only four case studies identified in the recently published works, there remains an overall research gap.

A major contribution in the field of EC implementation is the contribution by WÄNSTRÖM AND JONSSON [17]. In their research, they accompanied a materials planner (MP) during an EC occasion. Focusing on one specific EC, they observed how the MP manually controlled purchasing orders and lot sizes to reduce the obsolescence costs. With this manual control at the expense of higher administrative effort, the scrapping costs amounted to 3000€, low for a single change, but in the big picture of up to 1000 changes per month [11] this results in millions lost each year. As a key to increase efficiency and reduce costs, they concluded that high information quality (i.e., communication) was detrimental to the outcome, next to interchangeability of components. However, as they pointed out in a follow up research [26], it becomes evident that manual control and the administrative costs cause high strains on capacity, which they propose mitigating by prioritising on high impact components, effectively accepting higher scrapping costs. Hence, they introduced a method to develop a decision tree model and by applying a three-level filtering funnel on various logistic and EC variables, they created a company-specific decision model, resulting in a reduction of scrap costs.

In a recent addition, POTDAR AND JONNALAGEDDA [54] identified that a lack of proper communication is the main cause for problems within the implementation phase. Their work, based on experiences from the automotive industry, suggests improving the process flow and clarify interfaces between process partners. For that, a framework consisting of 22 steps was proposed with standardised steps and standardised language, with a database set-up to manage all occurring data and link to existing systems.

Following up on a previous literature review, IAKYMENKO ET AL. [6] conducted case studies to analyse the factors affecting EC implementation performance in Engineer to Order (ETO) firms. Although the investigation was limited to the Norwegian maritime industry, the 11 factors identified are transferable for any design and manufacturing project and impact the overall result of EC. Interestingly, the ranking of their factors seems to contradict previously published works, as communication is not directly included, but rather implicitly shown as a low-ranking factor of distance between project actors. Moreso, capacity is not even mentioned. As the highest-ranking factor time of EC occurrence has been indicated, which is typically addressed in ECM by front-loading methods. The other top factors were competence of the partner yards (external manufacturing) and the degree of vertical integration with suppliers. The difference to the study by WÄNSTRÖM AND JONSSON [17] might be a direct result of the focus on the ETO industry, as they state that in ETO industries, ECs are implemented immediately instead of being planned ahead. In related research [52], they point out that additional tools are required to improve efficiency.

The four case-study contributions presented, show that further investigative research in industry is essential, as the factors affecting EC implementation are highly dependent on the target industry and company. As main contributors in mass customisation industries, communication and capacity have been identified, as opposed to the occurrence and competence of partners in the ETO industry. In both industries, scrapping costs show to be the main driver of variable EC implementation costs which can be improved by proper EC implementation methods.

2.3.2 Engineering Change Implementation Modelling

The key challenge tackled by modelling research in EC implementation is the complexity and changefulness of the EC process. Publications have the main goal of understanding the process through simulation and reduction of complexity. To manage the inherent complexity, any applied modelling language and system requires adequate flexibility. Thus, LIU ET AL. [18] propose a flexible workflow system based on Petri-nets to simulate the complexity of the EC process. Their research targeted the process control itself, trying to set a dynamic base for future workflow management systems, giving them the ability to supervise complex processes such as EC. It was further followed up by LI ET AL. [59], who applied it to an aero-engine assembly.

In the field of new product development (NPD), EC implementation becomes especially complex and important. For instance, research concerning interdependencies between NPD and EC by LI AND MOON [60–62] simulated the effects of resource constraints and other factors on the overall lead time, with the capacity constraints being the main determining factor. In their contributions, they concluded that EC is an important aspect for successful NPD and provide a model to address company-dependent decision making suggestions. Though their model provides insight into NPD and EC, they suggest that further work in the area is needed to enhance the parameter setting and integrate company wide data to evaluate EC to achieve an optimal EC implementation process output,

A main strategy to mitigate the downsides of EC on NPD and manage complexity is the grouping of multiple changes into batches, described as a result of research by BHUIYAN ET AL. [63]. With their modelling approach, they were able to simulate different implementation scenarios, concluding in said batch introduction being superior, where multiple EC changes are introduced at a single point in time. However, due to immediate emerging problems, a mechanism should always be available to process urgent changes. An optimal batching frequency was identified by AHMAD ET AL. [64] through simulation, resulting in a project-specific optimum, through reduction of rework. Batching of EC is further supported by WYNN ET AL. [65], who used change propagation prediction through Monte Carlo simulation to better schedule change processes and reduce rework during the implementation phase.

To enhance communication, in an optimal EC implementation process, all process partners are involved as soon as possible but based on case-specific necessities. Out of the modelling approaches, it is evident that collaborative work and increased communication are key to optimal EC implementation, and an inclusion of downstream process partners in early design phases is pivotal in ECM. Several contributions [53, 66, 67] addressed this by discussing conflict management in concurrent engineering. Likewise to front-loading approaches, it was tried to analyse impacts on stakeholders through data dependency analysis, find optimal task-overlapping and in turn reduce the amount of rework during the EC design step, eventually reducing the time to implementation. Complementary research was done by REDDI AND MOON [68, 69], who through modelling and simulation showed that the optimal overlap between phases and tasks, as well as supplier involvement amongst others, is case sensitive. It was proven that every parameter under observation had its own optimum when isolated and the supplier competency in EC is key for Original Equipment Manufacturers (OEM) ECM performance, matching the factors identified in [6].

2.3.3 Implementation Strategies of Engineering Changes

Literature on implementation strategies remains consistent throughout the years, as few new publications emerged. The main effort lies in the reduction of implementation costs, such as obsolescence and rework. For instance, BARZIZZA ET AL. [23] followed up on the EC categories described in [56], adding their own taxonomy, and similarly to BALAKRISHNAN AND CHAKRAVARTY [51] calculated an optimal implementation date. Contrary to the latter work, they exclusively focused on direct part costs, such as rework, scrap, and quality. After conducting a case study, they conclude that to achieve an optimal state in EC implementation, exhaustive reporting and reduced time to implementation is required. Further, they suggest to dynamically calculate the optimal effectivity date, as the demand figures fluctuate on a daily base.

SHIAU [24] further researched an optimal effectivity date under aspects of production and engineering capacity, by listing all relevant data and combining it in a linear programming model on a manufacturing cell level. Their outcome shows a promising result in calculating the effectivity date, although the calculations took more than an hour for complex changes. Thus, they suggest investigating into heuristic approaches to reduce calculation time.

Other implementation strategy related papers offer qualitative approaches, for instance SCHUH ET AL. [15] and STEKOLSCHIK [27] cluster ECs based on their characteristics and provide strategies on how to implement the EC. Both, and other similar qualitative studies are provided as rigid deterministic systems, where only a certain type of change can occur, and a defined rule-set is used to implement. STEKOLSCHIK [27] for instance suggests focusing on high impact high effort ECs, and to omit controlling simpler changes.

Finally, some works are applied in earlier EC process steps, but directly impact the implementation phase. For instance, XUE ET AL. [70] designed a method, to consider and optimise design impacts (allocated in step three of the EC process) based on their effects on downstream (mainly implementation) processes. By employing the Taguchi method and associating noise parameters, they were able to lower the implementation cost by usage of robust design.

To summarise, EC implementation research is split into three distinct research streams and the research gap initially shown in [4] still remains valid, as no new publications have discussed the open fields such as scheduling optimisation or management of timing and deadlines. As seen in Table 2.1, authors either focused on the costs and finding an optimal date, on modelling the business process, or on finding factors affecting change implementation. Furthermore, the challenges in EC implementation seem to vary drastically depending on the industry observed. While research in EC implementation in mass-customisation as well as in ETO industries tries to reduce the costs of implementing changes, the actual steps taken differ gravely.

Table 2.1 Overview on EC Implementation Literature

Source	Type[a]	Goal	Solution
[64]	M	Change impact reduction	Optimal batching frequency
[23]	S	Implementation scenarios	Optimal date
[63]	M	Complexity control	Batching
[71]	M	Workflow development	Workflow development
[72]	CS	Implementation Factors	Process understanding
[6]	CS	Implementation Factors	Process understanding
[59]	M	Implementation Process Model	Petri-Nets
[60]	M	Decision Support	Process Simulation
[61]	M	Decision Support	Process Simulation
[62]	M	Decision Support	Process Simulation
[18]	M	Workflow Monitoring	Flexible workflow simulation
[67]	M	Workflow Optimisation	Simulation of Task overlapping
[53]	M	Change understanding	EC impact estimation
[66]	M	Workflow Optimisation	Simulation of Task overlapping
[73]	M	Workflow Optimisation	Simulation of Task overlapping
[54]	CS	Workflow Optimisation	22-step standardised process
[68]	M	Workflow Optimisation	Simulation of Task overlapping
[15]	S	Complexity Control	EC implementation strategy based on complexity
[24]	S	Implementation scenarios	Linear programming model
[74]	M	Workflow Optimisation	Modelling ideal EC implementation process
[27]	S	Capacity constrain mitigation	Prioritisation based on EC impact
[17]	CS	EC effects on material planning	Manual control of ECs
[26]	S	Capacity constraint mitigation	Prioritisation based on costs
[65]	M	Reduction of Rework	Simulation of change propagation
[70]	M	Change impact reduction	Simulation of change propagation

[a]M = Model, S = Strategy, CS = Case Study

Out of all papers identified, only one employs AI for dealing with the challenges of EC implementation. This may be due to several reasons, one being the complex and customised EC process itself, and only human capabilities providing the flexibility to control it. Four major challenges addressed in current literature have been identified, shown in the leftmost column in Table 2.2.

Table 2.2 Current challenges in EC implementation

Potential for improvement in EC implementation	Identified in EC implementation contribution
Change Timing	[17, 23, 24]
Process control	[60–62]
Scheduling control	[4]
Capacity & complexity management	[15, 26, 27]

These can in turn be transferred into four distinct BuNe:

(1) Change timing, discussed as the business need optimisation.
(2) Process control, discussed as decision support.
(3) Scheduling control, addressed by supervision.
(4) Capacity and complexity management, discussed as automation.

In an environment where every company needs to compete for personnel and material resources the application of AI could potentially support with these difficulties, even if it means to only automate controllable tasks such as supervision or supply ordering. Based on these four identified fields and global BuNe in ECM, the remainder of this thesis provides a knowledge base and design artificats for BuNe 1–3.

Publication I: AI-Artifacts in Engineering Change Management—A Systematic Literature Review

3

3.1 Abstract

Changes and modifications to existing products, known as engineering changes (EC), are common in complex product development. They require appropriate implementation planning and supervision to mitigate the economic downsides due to complexity. These tasks, however, take a high administrative toll on the organisation. In response, automation by computer tools has been suggested. Due to the underlying process complexity, the application of artificial intelligence (AI) is advised. To support research and development of new AI-artefacts for EC management (ECM), a knowledge base is required. Thus, this paper aims to gather insights from existing approaches and discover literature gaps by conducting a systematic literature review. 39 publications applying AI methods and algorithms in ECM were identified and subsequently discussed. The analysis shows that the methods vary and are mostly utilised for predicting change propagation and knowledge retrieval. The review's results suggest that AI in EC requires developing distributed AI systems to manage the ensuing complexity. Additionally, five concrete suggestions are presented as future research needs: Research on metaheuristics for optimising EC schedules, testing

---

Burggräf, P., Wagner, J., Saßmannshausen, T. et al. AI-artifacts in engineering change management—a systematic literature review. Res Eng Design ***35****, 215–237 (2024). https://doi.org/10.1007/s00163-023-00430-6*

O. Radišić-Aberger, *Optimisation and Control of Engineering Change Schedules in the Automotive Industry with Metaheuristics and Machine Learning*, Findings from Production Management Research,
https://doi.org/10.1007/978-3-658-51074-9_3

of stacked machine learning methods for process outcome prediction, establishment of process supervision, development of the mentioned distributed AI systems for automation, and validation with industry partners.

3.2 Introduction

Changes to products are a rule rather than an oddity [14]. Without innovating and adopting products, companies risk losing market share and revenue [14, 51]. These changes to the behaviour, function, or structure of a technical component are generally known as engineering changes (EC) [4]. With estimated costs per change accumulating to 20,000–50,000 € and an average of 1,000 changes per month at a German automotive OEM [11], the economic implications of EC and its management (ECM) are large. Failing to respond to the challenges has been attributed to costs as high as 30% of a component's price [6], a 20% increase in warranty charges [7], and an additional 15% in cost for redesign [14]. Within complex industries, such as aviation and automotive, the number of ECs have been shown to have increased in the past [40], a trend which is expected to continue with the shift toward highly iterative product development [2]. Since change is inevitable, its efficient handling is a key competitive edge.

However, the EC process is complex and versatile. The process flow is not linear, but iterative [12], with frequent loops and extensive communication in interdepartmental teams [75]. Additionally, a change to one component may induce changes to other components, spreading throughout the entire product. This phenomenon of change propagation [13] and the anticipation of the potential economic impact is at the core of ECM research [46]. Furthermore, from a supply chain perspective, even if an EC affects only one component, it can invoke an avalanche of logistical tasks [56]. As a result, contributions to EC and ECM focused on controlling the EC process and managing its complexity [4].

Another source of shortcomings for adequate ECM is limited capacity. Occurring ECs have to compete with regular tasks for resources [17], and thus, capacity is too limited to control all ECs adequately [26]. Up to 30% of an engineer's capacity [12] and 70% of manufacturing capacity are consumed by EC [76]. Furthermore, due to change propagation, it is near impossible for even experienced staff to anticipate all effects of an EC on the finished product [14]. The reduced predictability of EC effort is further exacerbated by fluctuations in supply chains [7] and the increasing complexity of products [15].

As a response, researchers suggested computational tools to support, anticipate, and automate EC related tasks. Barzizza et al. [23] for instance devised an optimisation method to calculate stock-optimal effectivity dates, whereas CLARKSON ET AL. [13] developed computational methods to predict change propagation paths. To augment computational ECM support, research suggests addressing two key challenges: Flexibility due to the complex nature of EC [18] and dynamic decision-making due to change propagation [77] and supply chain uncertainty [22]. These challenges for advanced computational support of the EC process can be translated to the ability of flexible, human-like decision-making, which is the scientific research area of artificial intelligence (AI). By applying AI, new possibilities to automate EC tasks arise [21], for instance, automatic design adaptations [20] and problem identification [78]. However, these AI-methods are not only capable of support for single tasks, but moreover increasingly able to manage and control complex processes [79]. Thus, targeting a future of cyber production systems [79], it is suggested to automate the entire EC process by researching new AI-artefacts.

Research of new AI-artefacts falls into the domain of information systems (IS), where design science research (DSR) has established itself as the predominant research paradigm [80]. To ensure scientific rigor and equally focus on practical appliance, the DSR framework by Hevner et al. [38] is commonly used as a guiding methodology [39, 81]. Hence, as IS and ECM research are both rooted in practical appliance requiring scientific rigor, we suggest applying the above DSR framework for future research on AI-artefacts for ECM. Accordingly, these new artefacts should be based on business needs (BuNe) from the environment, supported by a knowledge base from theoretical foundations and available methodologies within the domain. These four BuNes for ECM were defined by RADISIC- ABERGER ET AL. [1] as *automation, decision support, optimisation,* and *supervision.* Current literature reviews, however, have not covered how current AI-methods in ECM address these BuNes.

Thus, this study aims to establish a knowledge base for research and development of new AI-artefacts for EC control by systematically reviewing published AI-related contributions in the domain of ECM. As a review procedure, we apply the framework for IS literature reviews by VOM BROCKE ET AL. [82]. In this context, the necessary knowledge base is built on insights and identified research gaps and incorporates experiences, expertise, and existing artefacts from the application domain [83]. Hence, our review of AI-artefacts within the EC domain is driven by two main research questions (RQ):

- RQ1: What insights can be gathered from AI applications in the EC domain?
- RQ2: What future research is necessary for AI-supported ECM?

The remainder of the contribution is structured as follows: Section 3.3 defines the BuNes in the application environment of ECM and provides a review of existing literature. In Sect. 3.4, the systematic literature review method is described before discussing the results in Sect. 3.5. Section 0 answers RQ1 by providing the insights derived, which are the base for providing the research gaps and answering RQ2 in Sect. 3.7. Finally, a conclusion and outlook are provided in Sect. 3.8.

3.3 Background and Related Work

To answer the RQs and subsequently establish a knowledge base for developing future AI-artefacts for ECM, two background concepts are described which are later used for organising the literature: (1) the four BuNes identified for EC control and how they are currently addressed, and (2) related EC literature reviews and their framework.

3.3.1 Business Needs for AI Supported Processing of Engineering Changes

Although ECM research is driven by the five overarching goals of *less, more effective, more efficient, earlier, and better* [12], publications along the core EC process focus on earlier, more effective, and more efficient EC handling [4]. The goal of increased efficiency is commonly interpreted as the reduction of EC-related costs [26] and the best use of resources [52]. Effectiveness is improved by decision-making support, while earliness is improved by predicting problems before they occur. It is generally recommended to involve downstream process partners as soon as an EC is requested [53]. However, decision-making with partial knowledge, lengthy processes, and a multi-departmental environment without visible ownership cause ECs to become a bottleneck process instead of an enabler for product maturity [54]. Thus, Radisic- Aberger et al. [1] suggest research on new AI methods based on four identified BuNes which have been simplified to:

- BuNe 1: Automation
- BuNe 2: Decision support
- BuNe 3: Optimisation
- BuNe 4: Supervision

The first BuNe targets the main challenges of complexity control and capacity increase, which emerge due to the EC process itself. Past literature mainly resolved the issue with qualitative solutions. For instance, LIU ET AL. [18] conclude that the complexity of the EC process requires a flexible workflow system such as Petri-nets, while BHUIYAN ET AL. [63] have elaborated that introducing multiple ECs simultaneously as a 'batch' is superior to individual introduction. This batch idea was further discussed by WYNN ET AL. [65] and AHMAD ET AL. [64] who provide methods to identify optimal frequencies and batch sizes for EC scheduling.

The second BuNe has been addressed in several contributions by simulations or rule-based methods. WÄNSTRÖM ET AL. [26], for instance, suggested a decision tree model for application with various logistic EC variables. Similarly, LI AND MOON [62] provide a model to address company-dependent decision-making suggestions by integrating various data sources. OUERTANI [73] as well as REDDI AND MOON [68] show that predicting the impact of changes is effective for conflict resolution and effectiveness in EC.

BuNe 3 is more related to later phases of the EC process, with several publications concerning effectivity date optimisation, e.g., [23, 24] and optimal test schedules [84], but also optimisation of change propagation paths [85]. Although differing in their goal, the overall procedure is similar: An optimisation function is solved for a set of variables and based on the nature and different strategies based on the EC trigger suggested. Lastly, due to the EC process being lengthy and problems occurring unexpectedly [56], the entire process should be monitored dynamically resulting in BuNe 4.

3.3.2 Artificial Intelligence

AI is a vast research area, composed of six main disciplines along the total touring test: natural language processing (NLP), knowledge representation, automated reasoning, machine learning (ML), computer vision, and robotics [34]. Furthermore, the algorithms developed in AI research are often applied in other domains, which leads to confusion on what is and what is not AI [86]. For our review, we only consider methods as AI, if they fit into the definition given by RUSSELL AND NORVIG [34]: '*AI is the study of agents, that receive percepts from the environment and perform actions based on them*'. What characterises AI algorithms from simple input-output algorithms, is the ability to improve performance by receiving feedback on the actions they took towards a specific goal. We give a brief introduction of the algorithms in the discussion section for those algorithms, which are used in ECM. For a discussion of other AI methods, we refer to classical AI literature such as [36] and [34].

The application of algorithms is neither limited to their AI discipline nor the domain of AI. For instance, computer vision often utilises the ML algorithm convolutional neural networks, NLP approaches employ recurrent neural networks, and some algorithms find application in other domains such as data mining [34, 87] or dependency modelling [88]. Thus, AI algorithms are often not classified by their discipline, but rather by the task, they were initially developed for. For this, Russell and Norvig [34] classify tasks into five types: Problem-solving (PS); knowledge, reasoning, and planning (KRP); uncertain knowledge and reasoning (UKR); ML; and communication, perceiving, and acting (CPA).

3.3.3 Related Literature Reviews and the Holistic ECM Literature Framework

With the abundant amount of data recorded in the EC process, it becomes apparent that AI-supported handling of EC might help control complexity and mitigate the pains in ECM. However, although several literature reviews in the domain of EC exist, none focuses on AI applications. Whereas Jarratt et al. [43] focused more on standardising terminology, Hamraz et al. [4] sorted literature into a holistic framework (Fig. 3.1), which we are later adapting in Sect. 3.4.4 for classifying AI applications in the ECM domain. This framework by Hamraz et al. [4] is especially effective at identifying research gaps in literature, as it builds on the generalised EC process [46]. Within this framework, group ‘A’ consists of literature focusing on the pre-change phase, dedicated to minimising the number of changes. Group ‘B’ considers literature explaining and developing methods and tools to improve the in-change tasks and processes. Analogous to group ‘A’, group ‘C’ embodies research exploring the post-change stage of changes. The residual group ‘D’ covers publications on general topics and surveys in the field of ECM and related research areas. Additionally, Hamraz et al. [4] identified groups ‘B’, ‘C’, and sub-group ‘D2’ as the core areas of ECM research, with the others crossing domains with product design and project management.

Most literature reviews that followed afterward are based on one or both major publications described above. Of the seven literature reviews identified after [4], none discusses AI application in ECM extensively. While change propagation understanding and impact prediction are discussed in [76, 89, 90] as well as [91], [92] is a review on data usage in ECM. Literature reviews focusing on

EC implementation discuss the applicability of ECM research in engineer-to-order industries [58] or classify EC research towards production, product design, and logistics [93]. Among the reviews identified, [94] stands out as a meta-study toward standardising the language used in ECM. Finally, the review on change propagation analysis, its use-cases, and associated data representation by BRAHMA AND WYNN [16] touched on the application of AI-methods, although they did not discuss AI-methods for other EC tasks.

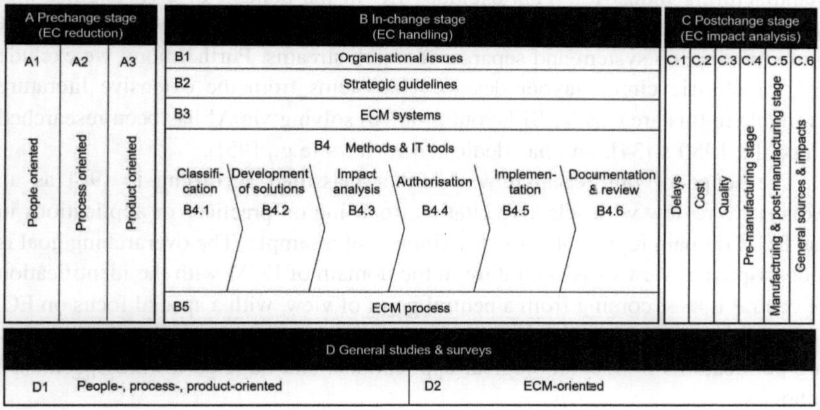

Fig. 3.1 The holistic ECM literature framework of HAMRAZ ET AL. [4]

In conclusion, although several literature reviews have been conducted in the past within ECM, none has widely discussed the application of AI. Hence, to close this research gap and provide a knowledge base for future AI-based research, we present the results of a literature review on AI applications in ECM.

3.4 Method and Materials

To achieve a reproducible result, a systematic literature review according to the IS literature review procedure by VOM BROCKE ET AL. [82] was conducted. The methodology is split into five steps as follows:

- Defining the review scope (cf. Sect. 3.4.1)
- Conceptualisation of topic (cf. Sect. 3.4.2)

- Literature search (cf. Sect. 3.4.3)
- Literature analysis and synthesis (cf. Sect. 3.5)
- Research agenda (cf. Sect. 3.6 and Sect. 3.7)

3.4.1 Review Scope

As our research domain, we focus on EC in the context of goods producing manufacturing industries. This excludes EC in the context of civil engineering, service industries, and software engineering, as these have essential differences in the production system and separate research streams. Furthermore, we exclude EC for electric circuit layout design. This stems from the extensive literature available in this area, as VLSI layout problem solving via AI has been researched since the 1980 s [34], and has dedicated reviews (e.g., [95]).

The scope of our research was further classified according to [96] as an exhaustive review with selective citation, focusing on practices or applications in the ECM domain represented by descriptions of a sample. The overarching goal is a descriptive review of AI literature in the domain of ECM with the identification of central issues, coming from a neutral point of view, with a special focus on EC. It is organised conceptually in a framework (Fig. 3.3, Fig. 3.4, Fig. 3.5), grouping publications by methodological applications, and targeted toward specialised scholars.

3.4.2 Conceptualisation of the Topic

A wide conceptualisation of the topic was done to get a broad understanding and an overview of current trends in the scientific community. This step was done by consulting standard literature in the field of ECM, such as books for basic definitions and other literature reviews to identify current ideas. Gaining a broad understanding and looking for synonyms of the topic, a concept map as suggested by Rowley and Slack [97] was drawn (Fig. 3.2).

With the key term of search string 1 (S1) being the target domain of 'EC', F1 represents the four BuNes introduced before and their respective synonyms. Likewise, F2 embodies the target solution principle of AI. Although Rowley and Slack [97] suggest applying different combinations of the search strings to discover literature, not all are sensible. Terms F1 and F2 combined, as well as on their own, would result in unspecific general literature outside the target domain, thus, these were omitted. Initially, it was intended to apply only the full combination of the three strings. However, during keyword search of the major databases 'Scopus', 'IEEE' and 'WebofScience', the results were comparatively low (<100 results for S1 + F1 + F2 and <250 for S1 + F2), which is why we settled on searching with only string S1 to make sure, that any publication within ECM applying AI-research would not evade the search due to inconsistent wording.

Hence, according to the methodology of [97], the search string was defined as S1: ("Engineering Change" OR "Design Change" OR "Product Change" OR "Engineering Design Management" OR "Technical Change"). The additional strings of F1 and F2 were then used as filtering criteria during manual screening of titles, abstracts, and the full text.

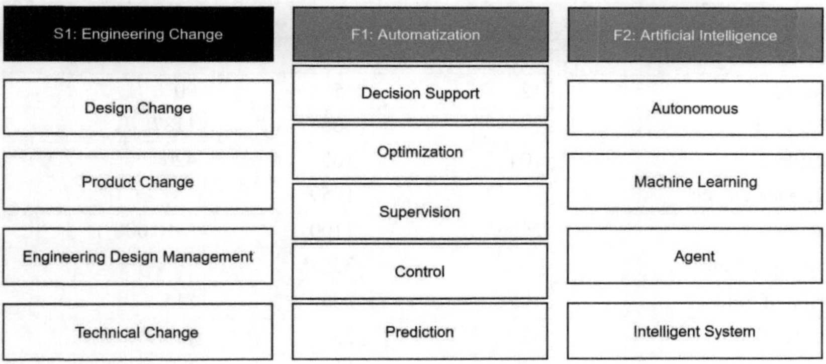

Fig. 3.2 Conceptualisation of the topic according to ROWLEY AND SLACK [97], with the main search string S1, and screening filter criteria F1 and F2

3.4.3 Literature Search

To identify relevant databases, research by GUSENBAUER AND HADDAWAY [98] was consulted. From their list of databases, we chose those that focus on 'Engineering', 'Computer Science', or 'Multidisciplinary', and have a Boolean search logic and a repeatable search. The result of the initial search is presented in Table 3.1. To evaluate the relevance of these contributions, quality criteria (QC) for inclusion and exclusion had to be defined. These are, according to the goal and scope of the review (Sect. 3.4.1), and the definition of AI given, as follows:

- QC1: Only journal and conference papers were included. This excludes books and practitioners' contributions, as the goal was to provide the state of art in academia.
- QC2: The publication must focus on EC as defined in Sect. 3.2 in the context of manufacturing industries (Sect. 3.4.1).
- QC3: To be relevant, the contribution must discuss the application of AI as defined in Sect. 3.3.2.

Table 3.1 Overview of search results per database of 26th and 27th March 2022

Search Strategy			
Data Base	Keyword	Title	Abstract
ACM Digital	65	77	277
Bielefeld Academic	378	1109	3969
DOAJ	12	15	90
EBSCO	77	306	1187
IEEE	104	105	420
Science Direct[a]		1859[a]	
Scopus	2606	1100	(>10,000[b])
Springer Link[a]		3558[a]	
Web of Science	185	205	815
Wiley Online	18	23	115
Total			18,675

[a]These data bases do not allow for a specific search strategy; thus, the total result is shown
[b]The result was omitted due to no feasible extraction method and no relevance to the topic of EC in random samples, and only the results from S1 + F1 + F2 and S1 + F2 were used

Due to language restrictions, only German and English literature is included.

Adhering to the PRISMA [99] flow chart (shown in Sect. 3.9, Figure 3.12), we filtered on relevant contributions for deeper investigation. Following the screening of titles and abstracts towards relevancy, 400 full-text entries were kept. After reading these 400 full-texts and applying the QCs, 28 publications on AI application in ECM remained and were considered for further discussion. To ensure we did not miss any publications, we further conducted a forward and backward citation search, based on the discovered 28 publications, as well as previous literature reviews. As a result, twelve publications were added to the review.

3.4.4 Organising the Literature

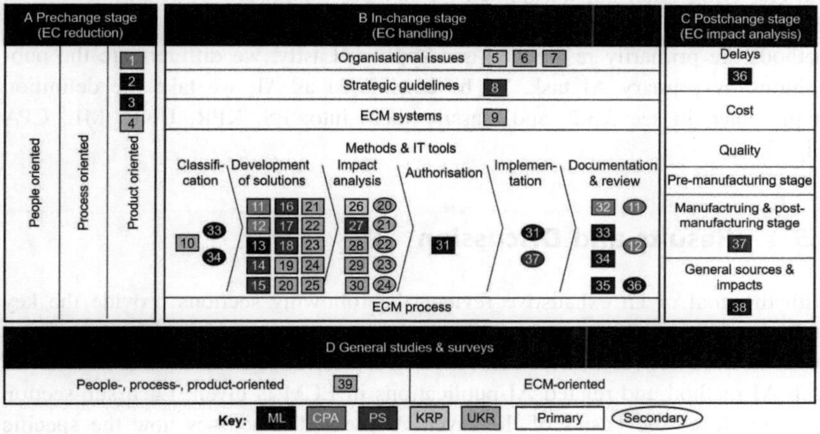

Fig. 3.3 AI-Literature in ECM clustered by application (primary and secondary) into framework HAMRAZ ET AL. [4] with additional layer of primary AI-Method employed. The numbers represent the second column in Table 3.2

Following the literature search, a three-criteria organising system as suggested by WEBSTER AND WATSON [100] was developed. The criteria (2) and (3) are used for quantitative discussion and criterion (1) for synthesis of future research needs:

(1) Primary BuNe addressed (Table 3.2, Figure 3.5)
(2) Primary EC research contribution (Fig. 3.3)
(3) Primary AI method applied (Fig. 3.3)

With the main goal given as establishing a knowledge base of AI-based applications for implementation of ECs, the presented four BuNes were defined as the primary criterion for identifying research gaps (Table 3.2, Section 3.7). As a second clustering criterion, we applied the holistic framework by HAMRAZ ET AL. [4] (cf. Section 3.3.3) to give an overview of where along the EC process AI methods are primarily researched and applied. Lastly, we differentiate the publications by primary AI task. To be considered as AI, we take the definition as presented in sec 3.3.2, and classify them into: PS, KPR, UKR, ML, CPA (Fig. 3.3).

3.5 Results and Discussion

With the goal of an exhaustive review, the following sections provide the key findings of the literature research. The first section presents a quantitative elaboration of the discovered literature, while in the following section a discussion of each AI method and related AI-publications in ECM is given. Each sub-section first introduces the basics of the given AI-method, discusses how the specific method was applied in ECM, and finally discusses each publication and their conclusions.

Afterward, form this analysis we draw the insights for RQ1 in Sect. 3.6 and correspondingly an answer for RQ2 in Sect. 3.7.

3.5.1 Overview of AI in ECM

Overall, AI-supported EC methods have been constantly researched, with most contributions being published after 2012 (Fig. 3.4). As seen in Fig. 3.3, the focus of general ECM research on the first half of the EC process is also present in AI-based literature, with most publications being published in the areas of solution development and impact analysis.

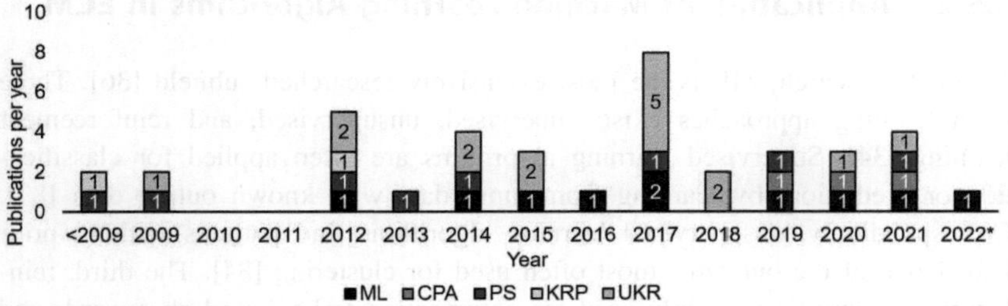

Fig. 3.4 Yearly development of AI based ECM literature

Of the identified contributions, 49% offer methods for decision support (BuNe 2), with the rest being split towards optimisation (BuNe 3) and automation (BuNe 1) (see Fig. 3.4). Literature focusing on this second BuNe employs various methods, though it UKR methods are dominant, whereas PS algorithms are mostly applied for optimisation algorithms. Interestingly, no publication discusses AI approaches to supervise the process (BuNe 4). This might be due to two reasons: Either the EC process is too complex for supervision by AI applications or the current state-of-the-art is satisfactory enough and there is no need for AI methods (Fig. 3.5).

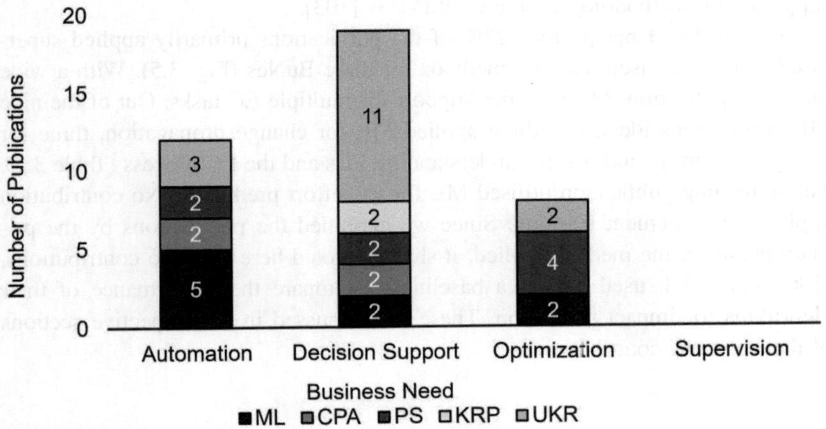

Fig. 3.5 Split of AI methods in discovered Literature vs. Business Need

3.5.2 Application of Machine Learning Algorithms in ECM

Of all AI research, ML is the most extensively researched subfield [36]. Three main learning approaches exist: supervised, unsupervised, and reinforcement learning [34]. Supervised learning algorithms are often applied for classification or prediction, by learning from input data with known output data [87]. Correspondingly, unsupervised learning algorithms find patterns without prior knowledge of the outcome, most often used for clustering [34]. The third, reinforcement learning is mainly used for optimisation tasks, based on rewards and punishment [87]. Additionally, ML algorithms are differentiated as shallow and deep learning algorithms based on the complexity of their deployment function [34]. Shallow learners require extensive manual data preprocessing [87], an application obstacle circumvented by introducing deep learning algorithms at the price of additional computational cost [34]. As the performance of the algorithms varies depending on the application [101], a lot of ML-based research in non-AI domains focuses on identifying the best algorithm and data preparation method for a given problem [86]. As with all algorithms, the performance of ML models varies with the preset hyperparameters. Thus, before final testing, hyperparameter optimisation runs should be conducted beforehand [102]. For evaluation which algorithm and hyperparameter set is best, usually a metric such as the F1-score or AUC-ROC (area under the receiver operating curve) is considered. In general, ML research follows the classic procedure depicted in Fig. 3.6a and Fig. 3.6b, supported by methodologies like CRISP-DM [103].

From an ECM perspective, 23% of the publications primarily applied supervised or unsupervised learning methods for three BuNes (Fig. 3.5). With a wide range of application, ML provides support for multiple EC tasks. Out of the nine ML publications identified, three applied ML for change propagation, three for change clustering, and two for understanding ECs and the EC process (Table 3.2). The remaining publication utilised ML for EC effort prediction. No contribution applied reinforcement learning. Since we classified the publications by the primary nature of the method applied, it shall be noted here that two contributions, [106] and [107], used ML as a baseline to compare the performance of their algorithms for impact prediction. These are discussed in the respective sections of their primary contribution.

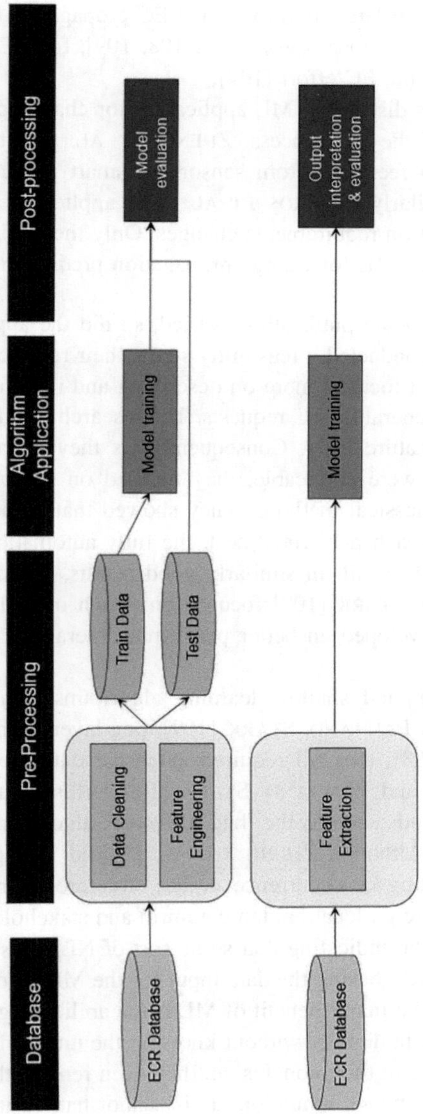

Fig. 3.6 Typical process for development of supervised ML (a, e.g., [104], unsupervised ML (b, e.g., [105])

3.5.2.1 Supervised Machine Learning Application

In ECM, supervised ML has been used to predict EC propagation based on information available from upstream processes [21, 108, 109], for EC understanding [110], and for predicting the EC effort [104].

The three publications discussing ML application for change propagation did so at different stages of the EC process. ZHENG ET AL. [21] tried to predict necessary ECs based on feedback from sensors of smart products, essentially triggering new ECs. Similarly, MORKOS ET AL. [108] applied an artificial neural net, to predict ECs based on requirements changes. Only the publication by PAN AND STARK [109] applied ML for change propagation prediction during the EC process.

As the overall focus of the publications varied, so did the application depth. As ZHENG ET AL. [21] conducted a feasibility study, their research did not compare to any baselines, and focused more on describing and investigating whether ML can be applied for generating EC requests. The research by MORKOS ET AL. [108] was on a more mature level. Consequently, as they already knew classical linguistic methods were applicable, they focused on comparing how ML performs against these classical methods. They showed that, although the semi-automatic linguistic approach performed best, the fully automatic application of artificial neural nets could result in similarly good results, without any manual input. Finally, PAN AND STARK [109] focused on which multi-label prediction method was best and developed an better performing hierarchical classification approach.

As all approaches applied shallow learning algorithms (random forest by ZHENG ET AL. [21] and PAN AND STARK [109], one-layer artificial neural net by MORKOS ET AL. [108], they all required extensive data preprocessing. As MORKOS ET AL. [108] and PAN AND STARK [109] utilised full EC requests as input data, NLP-methods such as the 'bag-of-words' algorithm were required for data preprocessing. Although ZHENG ET AL. [21]did not specify the data preprocessing, the resulting co-occurrence design structure matrix (DSM) was based on product sensor (e.g., location, temperature) and stakeholder (e.g., visual and natural language) data, indicating that some sort of NLP was applied. However, from an ML approach, basing the data input for the ML-algorithm on DSM seems counterintuitive, as a major benefit of ML is the ability to generate predictive models by feeding data directly without knowing the underlying dependency between the features [86]. In their conclusion, they even remark that basing their approach on a DSM is a major limitation, as it cannot handle incomplete case data, and suggest further investigation for actual data-driven ECM.

A second application of supervised ML was understanding the underlying factors of EC lead time. For that, SHARAFI ET AL. [110] performed an analysis of ten different algorithms within the supervised ML toolset to empirically derive the main drivers for long lead times. For that, they initially conducted an association rule analysis (a data mining algorithm), which was then used as input for their ML models. Of the 10 algorithms, none was performing significantly better at predicting the lead time. The initial goal of predicting the factors of long lead times was also somewhat unclearly shown. They argue, as the algorithms were able to predict accurately based on product complexity data, that these are factors for long lead times, without identifying their relative importance. This is somewhat at odds with the possibilities of ML. Although it is true that for most algorithms feature importance is not easily extracted, by repeatedly running the algorithms with and with certain features, their relative importance can be shown [111].

The third application of supervised ML in ECM was presented by RIESENER ET AL. [104], who applied a random forest algorithm for predicting the expected EC effort. They used real-world data to predict the effort needed to implement changes. However, they were only able to achieve a poor fit ($R^2 < 0{,}5$) and inaccurate predictions, concluding that random forest has its limits within EC and feature engineering is crucial. As this was also an exploratory study, they did not compare how other algorithms would perform. Furthermore, they have not conducted hyperparameter tuning, usually a very important step in ML model development. Thus, the outcome of their research can be only seen as a feasibility study.

Although all publications applying supervised ML conducted their research with real-world test sets, only PAN AND STARK [109] report on feedback from the organisation. Following the presentation of results, they report that the models should be tuned towards precision, rather than the F1 score, indicating that for their industry partner a correct classification is more important than finding all potential propagation paths. However, they too do not report on implementing the ML model into the productive environment.

3.5.2.2 Unsupervised Machine Learning Application

From the body of ECM research, four unsupervised ML approaches were discovered, focusing on clustering [105, 112, 113] and preparing data [114].

LI ET AL. [114] addressed the issue that DSMs are based on expert knowledge, which causes difficulties when applying them in productive environments. Thus, they designed a dendritic neural network to automatically generate a DSM based on the dendritic neuron model by TODO ET AL. [115]. Compared to the baseline, their algorithm did not perform better. However, they did show how to apply unsupervised ML to automatically generate DSMs for further application with change propagation methods. Interestingly, opposed to other ML publications, they based their research on a generic bicycle model, instead of a real-world problem. This is an immense benefit, as this enables other research to build on their outcome and compare their performance without requiring real-world data.

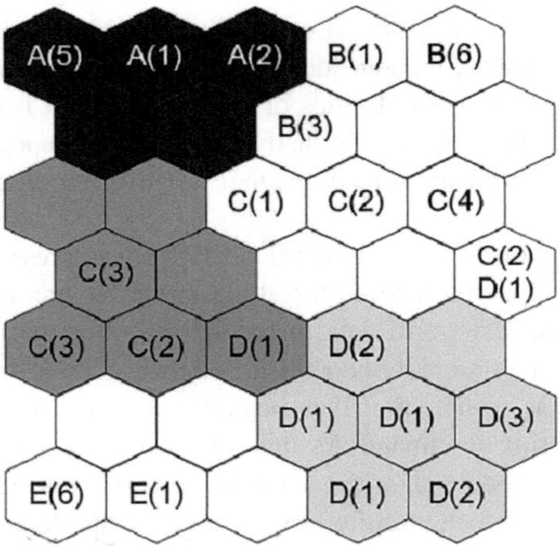

Fig. 3.7 Example of a self-organising map adapted from PACELLA ET AL. [105]. Colours represent clusters, letters the actual label, and numbers how many ECs are within that cluster

The final three applications of ML in ECM are applications for clustering and extracting knowledge through self-organising maps (SOM). Although all three applied SOM, their motivation varies. The research by GRIECO ET AL. [112] and PACELLA ET AL. [105] applied SOM for classification in case new ECs are drafted. For performance comparison, they evaluated the performance of the ML model using 54 ECs and compared the clustering result vs. expert knowledge (Fig. 3.7). They have shown that by applying the SOM method, they achieved an F1-score of 0.9 (with 1 being the maximum). However, they have failed reporting on the decision threshold, as well as conducting performance comparisons to other algorithms, thus their publication can also be classified as an explorative study. Note, that both contributions discuss the same research case, which is why there is no difference between the outcome of these two publications.

The research by Lu et al. [113] also applied self-organising maps, although for identification of anomalies in ECs and the EC process. Like the previous contribution, they utilised a real-world data set of 9849 ECs, and achieved a high accuracy rate of 96%. An important point they raised is the introduction of control and experimental groups for ML model development employing temporal split. In doing so, they ensured that no data leakage into train data occurs, and the performance during testing is comparable to deployment. However, they do not report on whether their approach was ultimately implemented in a productive environment.

3.5.3 Application of Communicating, Perceiving, and Acting AI-methods

The research field of CPA methods lies at hand: Retrieving information from data in the form of natural texts, images in large databases and learning machines to act upon these [34]. Within the field of ECM, the only primary CPA application is NLP (Table 3.2), with image processing being only remarked as a possibility by Zheng et al. [21]. NLP is the ability of a machine to interpret human language input and thereby retrieve knowledge from large data bases [34]. For training NLP models an extensive amount of data is required as the possible combination of letters to words are theoretically infinite. Furthermore, as human language is messy, data cleaning methods such as lemmatisation and stop word removal are applied during preprocessing steps [34] (Fig. 3.8). Once trained, NLP models can, among other applications, be applied for text classification, query response, and knowledge retrieval [34].

In ECM, four recent publications exist which have primarily applied NLP to advance ECM. The goal of the first three publications [116–118] is to retrieve knowledge from previous ECs, such that mistakes are not repeated. Contrary to their focus on retrieval from past ECs, Koh [119] applied NLP to evaluate social media comments, and derive future ECs from that. In their method, they follow the approach in Fig. 3.8.

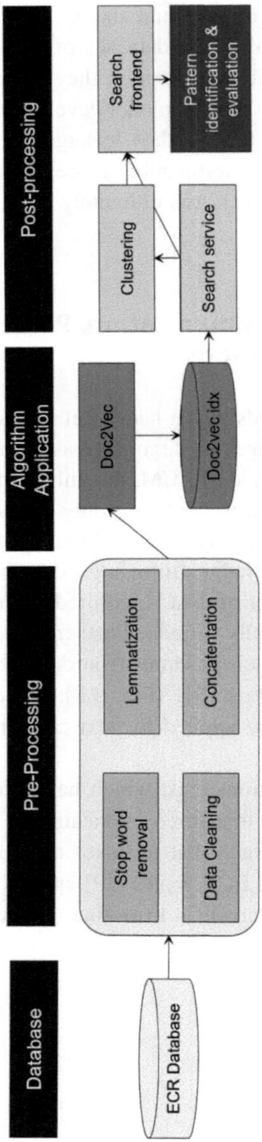

Fig. 3.8 Typical process for development of an NLP model application (e.g., [117])

While the first publication by ARNARSSON ET AL. [116] investigates whether a document embedding model outperforms an elastic search model, the other two apply a Latent Dirichlet Allocation algorithm to cluster ECs and provide support in knowledge management. Although all conclude that NLP is beneficial from an ECM perspective in retrieving information, they comment that the tailoring of the solution needs to be driven by the applying organisation. Furthermore, as the comparison in [116] shows, NLP outperforms elastic search only for longer search queries. These observations are attributed to highly individualised processes and language within companies. Additionally, due to applying the model to the EC database of a single company, the results cannot be generalised ARNARSSON ET AL. [116, 117].

As introduced, KOH [119] applied NLP for extensive mining of information from social media. In their research method, they do not vary so much from the ML approach by ZHENG ET AL. [21], although instead of applying the RF, they apply the classical CPA by CLARKSON ET AL. [13]. However, as they focus on the application of NLP, they describe the application of NLP in more depth and offer pseudo code. With their approach, they were able to analyse 3665 YouTube comments, pinpoint the component they affect, and derive ECs therefrom. Furthermore, they have shown that their method is not only applicable for the diesel engine they investigated, but also for 'Car', 'Vacuum Cleaner', and 'Washing Machine', suggesting it is applicable for all products.

Furthermore, we do note here that NLP methods are also important for data preparation for other AI-methods, as seen by the ML approach by PAN AND STARK [109] and MORKOS ET AL. [108].

3.5.4 Application of Problem-solving Algorithms

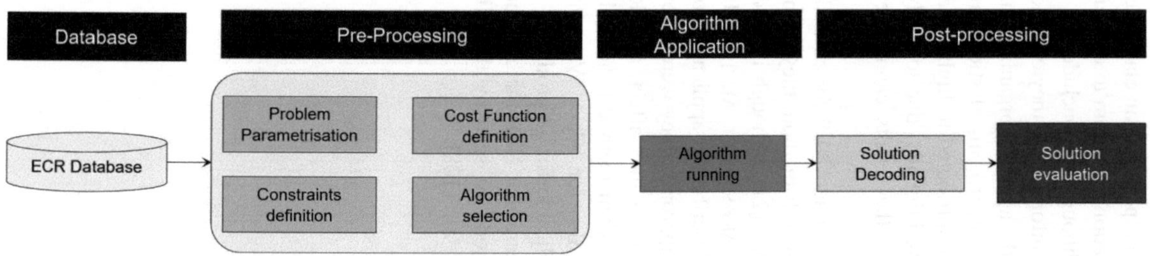

Fig. 3.9 Typical process for development of a metaheuristics model application (e.g. [20])

Problem solving via algorithms is a common subfield of operations research or computer science with countless exact solutions for some complex problems [25, 34]. However, AI research focuses on attempting to build machines that can find these solutions autonomously [34]. For that, AI research devised informed search algorithms known as metaheuristics (MH) [34]. Their main drawback is the inability of knowing whether the solution found is only a good solution or the best possible solution [87]. They are quick to adapt and find reasonable solutions even for complex combinatorial problems [25]. MH can also be combined with other AI-approaches. For instance, it is common to use GAs for hyperparameter optimisation or as layers in neural networks [87]. In principle, all MH application follows the same procedure (Fig. 3.9): Develop a mathematical model of the problem, run the algorithm for n number of iterations to find new possible solutions, evaluate the solutions, and accept the best discovered solution as the solution to the problem.

The application of MH in ECM is shared across all BuNes, although its frequent with optimisation tasks (Fig. 3.5). Two contributions for automation [20, 120] investigated how component design can be parameterised and subsequently design changes generated. Furthermore, MH were applied thrice for change propagation prediction [121–123], once for EC shop floor scheduling [124], once for impact prediction [106], and once for supply chain optimisation [125]. The most common algorithm applied was the genetic algorithm (GA), an easy to implement and flexible MH [20] (Table 3.2).

The two most notable contributions in this AI subfield in ECM utilise evolutionary inspired algorithms in feasibility studies for automatic design generation. As a study object, SHARP ET AL. [20] demonstrated how GA can be applied for automatic design optimisation after an EC is requested. Similarly, FATHIANATHAN ET AL. [120] applied a related evolutionary search algorithm to automate fixture changes after ECs. After defining the needs on how to approach the technical optimisation problem, both present a simulated case study. In the descriptive depth of their study, SHARP ET AL. [20] laid out a detailed automatic EC request workflow, depicted in Fig. 3.10. Comparatively, FATHIANATHAN ET AL. [120] focused more on arguing for and explaining the concept of evolutionary search algorithms. Both together have shown that MH are a possibility to automate parametric design, be it a discrete [120] or continuous [20] optimisation problem. SHARP ET AL. [20] further compared the performance of the MH vs. a human engineer, and although the calculation consumed more time, based on the input parameters the computer algorithm identified a better solution with less monetary resources invested, showing the potential of using AI to automate EC design activities. FATHIANATHAN ET AL. [120], however, has failed to show a comparison.

LI AND ZHAO [122] as well as LIAN ET AL. [123] employed MHs to identify change propagation paths for improved scheduling of solution development. Both approaches are not standalone applications of MH, but rather a combination of graph networks and MHs. With their research, they offer methods to reduce change propagation, and optimise the duration of the design phase of an EC. A key challenge faced by LI AND ZHAO [122] when applying the GA was the inability to a priori determine which design tasks are affected by an EC. As a result LIAN ET AL. [123], the model had to include all possible paths, irrelevant whether it will be followed, resulting in redundancy. In their follow-up research, LI ET AL. [121], they improved on this method by introducing more constraints, providing an optimal EC propagation path. The publication by considers using the improved Cuckoo search algorithm, instead of the GA. As with other publications, these three papers present their research based on an exemplary sample, but do not compare different algorithms or to other publications, thus they can also only be considered as feasibility studies.

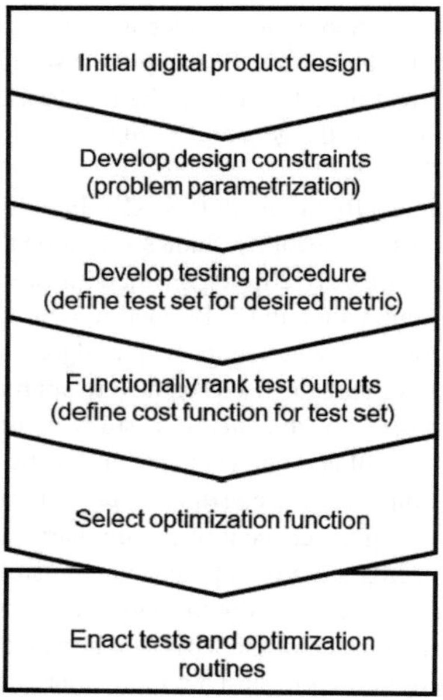

Fig. 3.10 Automatic engineering change request workflow as suggested by SHARP ET AL. [20]

Other MH research for capacity optimisation using a GA was done by WANG [124], who combined it with a Particle Swarm Optimiser (PSO) to minimise customisation costs by decreasing the job floor make span time of ECs by optimising manufacturing schedule. They showed that a hybrid solution provides better results than using only GA or only PSO, as it enables the algorithms to escape local optima. To further solidify their outcome, they compared the approach to an already established method, which performed worse than any MH application. Overall, their research offers a method to improve the EC realisation phase at the shop floor, an often overlooked area in ECM [4].

Impact prediction by application of MHs was shown by MEHTA ET AL. [106] who developed an own algorithm based on ant colony optimisation, a common MH. Their extensive research shows, how important attributes of ECs can be discovered by MHs and utilised for knowledge retrieval. In their research they compared their MH approach vs. ML models such as Decision Tree and Naïve Bayes classifiers. They have demonstrated that their approach is significantly better, although they do remark that their test set is comparatively small. This in turn leads to reduced interpretability and generalisation of their research outcome. At this point we note that their approach was further tested upon by CHEN ET AL. [107], which is discussed in the KRP section of our review, in accordance with its primary AI type.

Another MH application was suggested by WANG AND CHE [125] for preventive supplier selection concerning the complexities of EC. They applied four variations of the PSO to define the optimal product modification strategy by incorporating supplier data into the model. Not only did they mathematically optimise which EC to implement, but also provide a strategy on how to allocate call-off quantities among four suppliers. The extensive data used incorporated hard data as well as fuzzy data.

A common issue discovered in most PS approaches is the neglection of comparing the developed method with existing approaches and algorithms and testing with only few test cases. Except for the contributions by WANG [124] and MEHTA ET AL. [106], none has tested their approach against existing methods. Statistical significance was only proven in one contribution [106], although even here the authors mention that the test set size was too small. For those studies declared as exploratory studies this is acceptable to a certain degree, however, it does reduce the scientific impact of their outcome. Furthermore, no approach reported implementation in a running environment.

3.5.5 Application of Knowledge, Reasoning, and Planning Algorithms in ECM

The AI field of KRP centres on the ability of intelligent machines to store knowledge about their environment [34]. This subfield deals with logical representation of knowledge and lies the foundation of many advanced AI methods [87]. In ECM, logic can be applied for knowledge retrieval as shown by MEHTA ET AL. [126], or for scheduling optimisation by temporal logic as seen in [71].

In an initial publication, MEHTA ET AL. [126] apply KRP to predict the impact of new ECs by searching for similar ECs among historical data. In doing so, they provide decision support to avoid change effects before they are implemented into the product. This was done by the introduction of an own algorithm tested on several expert knowledge-based simulated data sets and compared to an available statistical approach and two classical approaches (metric space and set-theoretic). They show in their research, that EC similarity is indeed a good indicator for impact prediction, and in doing so, they set the foundations for their more advanced approaches (discussed in Sect. 3.5.4 and 3.5.6 according to the AI-method). Like their other publications, they compare to other algorithms and include statistical significance tests. However, by basing on a simulated test case, their publication shows only theoretical results. Furthermore, the test set ($n = 23$) is relatively small (17 train cases, 6 test cases).

Although situated in the domain of ECM, the temporal AI representation researched by CHINN AND MADEY [71] was a demonstrator for the AI-method itself in the wider area of workflow management. Nevertheless, their research shows the possibility to improve the workflow of EC during the design phase by scheduling ECs and associated tasks based on urgency, feeding to BuNe3. However, their research was an exploratory study and mainly contributed possible new research paths to incorporate temporal relations in scheduling problems. As such, no implementation in practice is reported.

3.5.6 Application of Uncertain Knowledge and Reasoning Approaches in ECM

The concept of knowledge representation shown in Sect. 3.5.5 mainly consisted of Boolean logic, where the outcome of an agent's actions are deterministic [34]. However, this is not necessarily always the case [87]. In some cases, the outcome of a process cannot be definitely precalculated, but only a probability given of a certain outcome [36]. Designing agents that can deal with this uncertainty is the AI-subfield of UKR. From our literature search, we have identified four approaches to deal with uncertainty in the ECM process: Probabilistic knowledge representation, Fuzzy logic, Bayesian networks, and distributed artificial networks.

3.5.6.1 Probabilistic Knowledge Representation

We identified two publications integrating probabilistic knowledge in ECM, specifically for impact prediction, the publication by MEHTA ET AL. [127] and the subsequent research by CHEN ET AL. [107].

Both contributions apply probabilistic knowledge representation to predict the impact of an EC. Their research shows, that similarity of change features of the EC are indicators for predicting the impact of new ECs. To proof, they simulated a test set of 23 ECs, of which a random combination of 17 was used as a knowledge base to predict the impact of the remaining 6. Additionally, as with the previously discussed approach by MEHTA ET AL. [126], the size of the test case is too small, which was acknowledged by CHEN ET AL. [107]. Furthermore, their test case is based on simulated data, thus it is not shown that the approaches will also work in a running productive environment. However, this is one of the rare cases identified in AI-based ECM research, where a later contribution [107] compares their performance against a previous contribution [127].

3.5.6.2 Bayesian Networks

As seen in [122], modelling every possible propagation path is a memory consuming inefficient method. Thus, AI-research has devised Bayesian networks (BN) as a prominent AI method to reduce memory and computation resources [87]. Closely related to probabilistic knowledge representation, BN are a systematic way to represent probabilistic relationships. They are directed acyclic graphs, where each node connects parents and children with a conditional probability distribution [34] (Fig. 3.11).

In ECM, BNs are generally used for impact and change propagation analysis to provide decision support (Table 3.2). In our literature search, we identified five publications applying BNs which are exclusively clustered into the third EC process step, impact analysis. Most research is motivated by improving from the current status quo of the change prediction method (CPM) and DSM. For this, LEE AND HONG [128] and YEASIN ET AL. [77] present a dynamic BN (Fig. 3.11) while MIRDAMADI ET AL. [129] designed a Bayesian belief network. Additionally, DIALLO AND ZOLGHADRI [130] provide a concept for a causal Bayesian network. However, only the two dynamic BN were tested on a data set against the classic CPM. The main benefits of change propagation prediction by BNs were reported to be additional flexibility and, in the case of dynamic BNs, incorporating time effects. Unfortunately, later research has not compared the performance to previous BNs. Furthermore, DIALLO AND ZOLGHADRI [130] propose hybrid networks to manage discrete and continuous variables.

Within the discovered publications, two more BN applications have been discovered outside the core EC process. DAMAK ET AL. [131] applies BNs to predict what type of EC an operational change would trigger. This is similar how requirements changes would trigger an EC and the application by MORKOS ET AL. [108]. Thus, their approach is more suited for estimation of effects before an EC is initiated, rather than to support the EC process.

Another goal of EC research and associated field is the prevention of changes [12]. One way of achieving this is by designing components in way, that they absorb changes [14]. To determine which components should ideally be flexible, HU AND CARDIN [132] develop a BN-based method. Although they show that it is possible to determine which components should have increased flexibility to absorb potential future requirement changes, their approach cannot address the cyclic dependency. Thus, they suggest dynamic BN for change propagation prediction, as shown by the publications presented above.

Change Propagation Probability			
	A	B	C
A	1	0.3	
B	0.5	2	
C		0.3	3

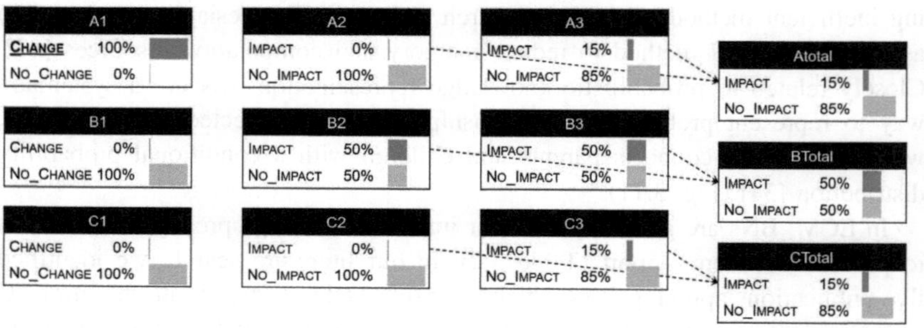

Fig. 3.11 Example of the change propagation model based on dynamic Bayesian networks by YEASIN ET AL. [77]. The table above shows the corresponding DSM

3.5.6.3 Fuzzy Logic

An early approach for AI were rule-based logic methods [34, 36]. However, one downside of these systems is that data is usually not available as a Boolean. Hence, fuzzy logic has been devised [34], and although it is today mostly researched as part of control systems research [87], for completeness we include it as an AI-method applied in ECM.

In ECM, fuzzy logic has been applied for instance to translate requirements by AGUWA ET AL. [133], workload assigning [134], and effort optimisation [135].

A difficulty when designing new ECs is the prioritisation of work, as it is rarely known which EC is most important [26]. This becomes even more difficult when ECs are triggered by quality issues [133]. In these cases, the warranty costs per case and the relative occurrence of these indicates a priority. However, interpreting and setting a rule for that requires fuzzy logic, as shown by AGUWA ET AL. [133].

A standalone method for automatically assigning ECs to the correct change manager was developed by SANDKUHL ET AL. [134]. In their approach, they constructed weighted graphs based on fuzzy string comparison to cluster ECs and the respective user. Although their approach is valuable, they mention as their main limitation that it was not validated with a running real-world case.

Another publication considered modelling the product development process to optimise the required effort [135]. To do so, they translated errors occurring during development into a fuzzy expert-based model and simulated how the effort is affected by product maturity. They showed, that while software tests are more beneficial in early project phases, hardware tests require a minimum project maturity to justify their effort.

3.5.6.4 Multi-Agent Systems

For some complex problems, such as the EC process, the actions of a single agent can depend on the actions of another agent within the same environment [34], or it is simply easier to model effects [136]. These systems are known as multi-agent systems (MAS), falling under the subfield of distributed AI systems [136]. A key concept of MAS, is the target for each agent to improve its performance measure, which is not necessarily a shared goal [87].

Within the body of knowledge, five agent-based AI applications were discovered, of which three are utilised to automate EC tasks [78, 137, 138], one to simulate change propagation [139], and an design evaluation [140] (Table 3.2). The motivation for all agent-based approaches is it to enable the researchers to connect multiple knowledge bases while retaining retrieval speed and manage EC complexity. In their realisation, each MAS is highly individualised, which is why these five publications are introduced individually.

CAMARILLO ET AL. [78] developed an MAS for problem-solving from a product quality perspective. By applying a domain-independent agent architecture, they digitalised the existing process at the industry partner for their agent system. Through adding an agent for each specific task and machine they were able to manage the complexity resulting from the manufacturing process. The system was subsequently tested and implemented at an energy storage production facility, and it showed that 80% of the solutions were sensible, and for another 10% the discovered proposal led to a better solution. One of the few publications discussing the implementation of the prototype in a running environment, they state that the availability of proper product lifecycle management and manufacturing problem solving is paramount.

Finally, for data handling automation BENDER ET AL. [137] compared four agent architectures for combining product and logistics-related data for more efficiency in product life cycle management systems. Their publication remained a concept, although it shows how specific tasks in ECM can be outsourced to the machine, and offer the architecture for other, similar application.

In some cases, an EC is not a new design, but a recombination of already available components. For this case, BEROULE ET AL. [138] devised a consensus seeking MAS of distributed fuzzy agents to integrate multiple departments and generate new combinations of components to resolve EC requests. They show the performance of this method by a simulated example of an office chair. However, they do not compare their performance to a human engineer or other algorithms. Rather, they just state that it is a possibility for automation.

Another MAS application for automation is presented by HABHOUBA ET AL. [140]. Having identified that ECM data is too varied to be processed by a single entity, they suggest dividing EC process into its subtasks, and devising an agent for every single one. They present a simulated case study of an airplane wing EC, showing the inner workings of the MAS. The final decision of the design however still lies with the human expert. From an application perspective, the split into multiple agents has a second advantage, overlooked by the researchers. With distributing the task into many sub agents, maintainability of such a system would be increased as the improvement of each agent could be done independently of the others.

Adding to BuNe2, MA ET AL. [139] apply an MAS to model and simulate the effects of ECs and provide decision support. In their work, they described ECs on components as agents, traversing a graph and simulated how changes to a component affect other components. A key component of their method is the inclusion of concurrent ECs, where multiple ECs are triggered simultaneously. In such a case (as seen with the Dynamic BN), ECs on other components have cyclic effects on other components. Furthermore, they increase the level of detail of the propagation analysis by basing their network on parameters and properties of components. Finally, they test their method on the case study of a gear box.

Overall, as all contributions were successful in their respective tasks, they showed that MAS are suitable for computational support of ECM. However, except for the contribution by CAMARILLO ET AL. [78], none of the proposals has reported a deployment into productive environments. Furthermore, it is interesting that no contribution has devised an MAS over the entire EC process. This could relate to the complexity of the EC process, as by Ashby's law [19] to fully grasp it, an equally complex MAS would be required.

Table 3.2 Overview on Artificial Intelligence Applications in ECM

Source	#	Type[a]	Subfield	Algorithm applied	Goal	Business Need
[119]	1	CPA	NLP	RoBERTa	EC anticipation	Automation
[21]	2	ML	Supervised ML	Random Forest	EC anticipation	Automation
[108]	3	ML	Supervised ML	Artificial Neural Net	EC anticipation	Automation
[132]	4	UKR	Bayesian Network	Bayesian Network	Requirements Propagation	Decision Support
[71]	5	KRP	Temporal Logic	Temporal AI	Workflow Optimisation	Decision Support
[138]	6	UKR	Multi-Agent Systems	MAS + Fuzzy Logic	Workflow Optimisation	Optimisation
[134]	7	UKR	Fuzzy Logic	Fuzzy string comparison	Workflow Optimisation	Decision Support
[125]	8	PS	Metaheuristic	Particle Swarm Optimisation	Supply chain optimisation	Decision Support
[137]	9	UKR	Multi-Agent Systems	Multi-Agent System	Workflow Automation	Automation
[133]	10	UKR	Fuzzy Logic	Fuzzy Logic	EC prioritisation	Decision Support
[116]	11	CPA	NLP	Document Embedding	Knowledge retrieval	Decision Support
[117]	12	CPA	NLP	Latent Dirichlet	Knowledge retrieval	Decision Support
[109]	13	ML	Supervised ML	Random Forest	EC detection	Decision Support
[120]	14	PS	Metaheuristic	Evolutionary Search	Design generation	Automation

(continued)

Table 3.2 (continued)

Source	#	Type[a]	Subfield	Algorithm applied	Goal	Business Need
[20]	15	PS	Metaheuristic	Genetic Algorithm	Design generation	Automation
[122]	16	PS	Metaheuristic	Genetic Algorithm	Change propagation	Optimisation
[121]	17	PS	Metaheuristic	Genetic Algorithm	Change propagation	Optimisation
[123]	18	PS	Metaheuristic	Cuckoo Search	Change propagation	Optimisation
[78]	19	UKR	Multi-Agent Systems	Multi-Agents System + CBR	Solution finding	Automation
[140]	20	UKR	Multi-Agent Systems	Multi-Agents System	Design evaluation	Automation
[128]	21	UKR	Bayesian Network	Dynamic Bayesian Network	Change propagation	Decision Support
[139]	22	UKR	Multi-Agent Systems	Multi-Agent System	Change propagation	Decision Support
[129]	23	UKR	Bayesian Network	Bayesian Network	Change propagation	Decision Support
[77]	24	UKR	Bayesian Network	Dynamic B. Network	Change propagation	Decision Support
[130]	25	UKR	Bayesian Network	Causal Bayesian Network	Change propagation	Decision Support
[127]	26	KRP	First Order Logic	Own Algorithm vs. Statistical	Impact Prediction	Decision Support
[106]	27	PS	Metaheuristic	Own Algorithm vs. Decision Tree	Impact Prediction	Decision Support
[107]	28	UKR	Probabilistic Knowledge	Own Algorithm vs. KNN	Impact Prediction	Decision Support
[135]	29	UKR	Fuzzy Logic	Fuzzy Logic	Impact prediction	Decision Support
[126]	30	UKR	Prob. Knowledge	Own Algorithm vs. KNN & SDA	Impact Prediction	Decision Support
[104]	31	ML	Supervised ML	Random Forest	EC Effort prediction	Decision Support

(continued)

Table 3.2 (continued)

Source	#	Type[a]	Subfield	Algorithm applied	Goal	Business Need
[118]	32	CPA	NLP	Latent Dirichlet	Change Clustering	Automation
[112]	33	ML	Unsupervised ML	Self Organising Map	Change Clustering	Automation
[114]	34	ML	Unsupervised ML	Dendritic Neural Net	Data preparation	Automation
[105]	35	ML	Unsupervised ML	Self Organising Map	Change Clustering	Automation
[110]	36	ML	Supervised ML	10 Algorithms	EC understanding	Optimisation
[124]	37	PS	Metaheuristic	PSO + Genetic Algorithm	EC scheduling	Optimisation
[113]	38	ML	Unsupervised ML	Self Organising Map	EC understanding	Optimisation
[131]	39	UKR	Bayesian Network	Bayesian Network	Flexible Design	Optimisation

[a] AI method and research field; ML = Machine Learning, CPA = Communicating, Perceiving, and Acting, PS = Problem Solving, KRP = Knowledge, Reasoning and Planning, UKR = Uncertain knowledge and reasoning

3.6 Insights for Future AI-application Development in ECM

Insight 1 (I1): Research on AI application in ECM focuses mostly on pre-approval and post-implementation process steps (Fig. 3.3).

AI-artefacts are existent in the field of ECM, although their application mostly focuses on the same areas as general ECM research. There is a high potential for AI application in ECM, going as far as to automate the entire EC process as suggested by some researchers. The main application target is to provide decision support, with a secondary goal to solve capacity constraints or find the optimal design change. The classification of ECs, although only once explicitly mentioned as a goal of AI-research, is often a byproduct of EC impact prediction. This leads to another key takeaway considering the application of AI-methods. The framework chosen shows the primary point of application in the EC process. However, as shown with the discussion of NLP and ML papers, for AI-application multiple process steps must be developed upon. For supervised learning this becomes especially apparent. Without existing labels, no ML-model can be trained. Thus, the last step, documentation and review is an important input parameter for any supervised ML application in ECM (e.g., [104, 109]).

Insight 2 (I2): Population-based MH and ML algorithms are versatile AI-algorithms for optimising and automating tasks during the EC process (Fig. 3.5).
This becomes evident when the range of the applications is considered. Most other AI methods were confined to one or two process steps, whereas ML algorithms cover most process steps. From a usual ML application perspective, it is interesting, that the first two process steps are neglected. As seen in Fig. 3.3, no publication researched exclusively ML-driven classification of ECs, although it is one of the most common ML applications. This is a direct result of our clustering framework, as only the primary contribution has been considered. Most publications do classify past changes, but mostly as a preprocessing step for other applications, e.g., change propagation. Regarding the development of solutions, we infer that this is due research occurring in other domains such as design automation, e.g., [141, 142].

Comparably, although MH approaches are concentrated on optimisation, they cover diverse tasks within this process step and are often a preprocessing step for ML application (e.g., hyperparameter optimisation). The limiting factor of these algorithms is their design—they always require a cost function, which they optimise their problem with. As such, only those tasks can be automated by MH, where variables can be parameterised, and an optimisation goal is given.

Insight 3 (I3): Multi-agent systems are required to fully automate the EC process as the data flow occurring in ECM is too decentralised and the process too complex.
Since research on AI-artefacts mostly focuses on single tasks during the process, full automation can only be achieved by linking them sequentially. However, due to high complexity and EC data coming from various IT-systems and different departments, a specialised solution is required for automating the EC process. In AI-based ECM research, managing complexity is solved by the introduction of MAS. By employing multiple smart agents, data management and maintainability are simplified, and the requirements of distributed stakeholders are fulfilled.

Insight 4 (I4): Within ECM, no standard data set exists to test new algorithms and most algorithm testing is performed on real-world data.
Most publications that tested their approach did so on real-world data. However, as MEHTA ET AL. [106] point out, to generalise, research on new tools should be done with standardised data sets, which are currently missing in the EC domain. Thus, the results derived from experiments on one data set are confined to the specific case-study, which in turn leads to the necessity to retest algorithms for each company and industry. Furthermore, as these data sets would have a lot of company specific information, they are probably subject to company secrecy. However, generating a simulated case as a reference set is a double-edged sword. With the simulation, possible anomalies existing within the EC process could be overlooked, or the problem too simplified, leading to wrong assumption on the performance of algorithms.

Insight 5 (I5): Although most AI-research is based on case-studies, only a few have reported implementation in the industry.
From all 39 papers reviewed, only one [78] implemented the prototype in a productive environment, and one [109] discussed algorithms tuning with an industry partner. The reasons for this can be manifold. One major difficulty might be the high individualisation of ECM or an insufficient data backbone for AI application. Other potential challenges are the necessity of consistent data management (e.g., product life cycle management systems) and data access [78]. As a result, AI-artefacts are possibly not yet adapted to usage in productive environments, or, at worst, the applicability of the research outcome is not given for productive environments.

Insight 6 (I6): The EC process is driven by expert knowledge and highly individualised, requiring extensive data engineering.
As shown by some of the ML methods, the input given by EC requests is often given in natural language. This however leads to difficulties. Not only does the language within ECM change in every organisation [117], but also the EC process itself varies. Thus, any application of AI would have to be tailored to the target organisation. The data engineering and mining could further be aided with statistical methods, not necessarily requiring expensive AI-methods.

Insight 7 (I7): EC data is mostly vague and the outcome of the process often uncertain, necessitating fuzzy logic and probabilistic methods.
The outcome of any process in EC, be it design generation, change propagation or any other, is difficult to grasp. This is also indicated by the lack of supervising

methods (BuNe 4). This can either be due to the unavailability of performance indicators, or due to inept methods. Any supervision method for the EC process, however, would have to manage this. This necessitates monitoring with uncertainty and probability.

Insight 8 (I8): AI based ECM Research has seldomly compared their approach to already existing solutions, showing overall lack of rigor.
A common theme observed was the introduction of new approaches and methods and claiming that they perform well. However, they have rarely compared their approach to solutions discovered in previous publications. This can be a direct result of the high individualisation of the EC process (I6), but also due to low rigor in AI-based ECM research. Furthermore, no publication has linked their actual code or algorithm, thus making it difficult for others to replicate the experiments on other test sets. Another general point of critique is the application of AI-methods without parameter optimisation. Parameters have a significant impact on nearly every AI-method, and should always be considered [102].

In conclusion, and as an answer to RQ1, the findings from the insights can be described as follows: The complexity and expert-knowledge-based EC process requires AI methods to be applied in ensemble when targeting automatic EC processing. As most ECs have similarly occurred in the past, using knowledge of past EC occurrences is suggested, although extensive data engineering is necessary. Additionally, we heavily suggest testing new methods against already existing approaches, to determine which method is superior.

3.7 Implications for Future Research

Following the discussion of results in Sect. 3.5 and insights in Sect. 3.6, this section is aimed at synthesising the outcome and answering RQ2. Resulting from (I1) we have identified that the research gap for EC implementation [4, 58] also translates to AI-driven ECM research. Looking at the literature from the perspective of RQ2 and the BuNes, we conclude that AI-driven ECM is possible, although numerous research gaps exist.

From the perspective of readily available tools, BuNe 2 and BuNe 3 are best covered. For instance, by adapting the clustering approaches of current AI applications towards the first step of BuNe 3 (Optimisation). Further useful methods are the decision support applications defined. As researchers were able to predict the outcome and impact of a change by the usage of EC data (e.g., [104]), the methods could be tested for application during EC implementation. However, this requires extensive testing and data engineering (I4), as the logistical and production data enriches the process. Because of unavailable standard data sets, comparing different approaches is impossible. This can have multiple reason, such as company policy and data complexity. However, this can be mitigated by sharing the algorithm and code. In doing so, other researchers can test the method with their own data set and compare the performance and improve upon these.

For BuNe 1, automation of the EC process, research, and development of a dedicated MAS would be necessary (I3). As even more stakeholders are involved during the EC process than during the generation of the design change, the ensuing AI application and agents would have to control various heterogenous aspects.

Hence, an initial start would be to automate those tasks, which are less communication-intensive, like the optimisation of the effectivity date (BuNe 3). This could, in an MAS, be easily done by MH or ML models (I2), while the ordering task is still performed by a human operator. Research would be required on incorporating multiple data sources into the optimisation models, as well as defining which MH algorithm is best suited. Likewise, classification and impact analysis, could be supported by ML method (I2).

From within our literature review, no direct answer or approach for BuNe 4 can be derived. However, with (I4) and (I7) in mind, the dimension of the EC process leads to data and information loss on interfaces, causing instability in downstream process steps. Thus, a research gap towards supervision techniques tailored to the EC process becomes evident. Additionally, while other fields of research have explored data needs for AI applications, EC data remains mostly vague (I7). Key challenges remain in mining knowledge from the process and joining various data types together for better process understanding. Future research should therefore also focus on establishing standardised data formatting and data sets, to test newly developed algorithms and methods.

A remaining research gap is a low report on the actual application in the industry (I5). Thus, we suggest conducting case-studies, whether AI-driven ECM has been implemented in various industries. This would further help understand the hindrances and difficulties. Furthermore, without validation in the field, AI-applications for research cannot confirm whether their approach is suitable for advances in ECM. Furthermore, several researchers developed new methods, however they have not compared their performance against previously published solutions (I8). This is insofar an issue, as researchers and practitioners alike cannot confirm, which method is best to be applied in industry or case-studies.

Table 3.3 Research gaps of business needs

Business Need	Research Gap	Transferable knowledge from other AI applications in ECM contribution
Business need 1: Automation	Modelling and development of a multi-agent system	[20, 78, 137]
Business need 2: Decision support	Decision support by usage of objective data, unsupervised ML to automatically cluster changes into similar batches	[104, 105, 110]
Business need 3: Optimisation	Metaheuristics approach to define optimal effectivity date	[20, 124, 125]
Business need 4: Supervision	Constant prediction of outcome of an EC scenario	–
–	Validation in the field by case study, and comparison of different approaches	[78, 107, 127]

Hence, aiming at RQ2, we can summarise the research needs as follows: While BuNe 1 requires research on modelling and developing an MAS, BuNe 2 and 3 have been already addressed partly in literature. Here, research should focus on determining the relevant factors, and testing various algorithms to determine the best outcome. Eventually, BuNe 4 would need exploration on AI-driven control of knowledge-intensive processes. To give an overview for future research and the literature gaps, Table 3.3 presents the BuNe, how they are currently addressed in the literature, as well as knowledge transfer possibilities from within the domain.

3.8 Conclusion

Management of ECs increasingly burdensome on design, logistics and production departments. Driven by this industrial need, automation by AI has been suggested by various authors. To aid the development of these new AI applications, we conducted a systematic literature review of publications on AI-artefacts in ECM. By discussing 39 identified publications, eight insights and five research needs were documented, thus establishing a knowledge base for future development of AI methods.

In summary, the potential for usage of AI in EC has been recognised by various authors. The most common application is to understand change propagation and change impact. This indicates that most of the problems in ECM emerge from this area. It has been established that conventional approaches are not providing satisfactory flexibility and dynamicity in the EC domain. More so, it is necessary to use different methods from the entire AI toolset, such as NLP, BN, OR, and ML in ensemble for a fully automatic data-driven EC process. Hence, for full automation of the EC process, an MAS is required.

The two RQs, “What insights can be gathered from AI applications in the EC domain” and “What future research is necessary for AI-supported implementation of ECs”, were answered by presenting eight insights and five research needs respectively. From the eight insights, in combination with predefined BuNe, future research on AI-supported ECM should focus on:

(1) Testing of various metaheuristics in the context of EC effectivity date optimisation
(2) Testing different combinations of ML methods to predict the EC effort.
(3) Researching distributed AI systems for automated EC processing
(4) Development of domain-specific process monitoring methods
(5) Validating AI-artefacts with industry partners

With this outcome the review provides two contributions to practitioners and researchers alike: Through providing a detailed analysis of current AI research in ECM, both gain an understanding of the challenges and possibilities already existing in the domain. With a discussion of these insights towards the BuNe we point towards literature gaps and opportunities for future research in ECM.

As a limitation, we point out that we have restricted the EC definition to technical artefacts in mechanical design and excluded AI-supported EC in areas such as construction engineering, software engineering or circuit engineering. Another limitation of our study is the provision of BuNe. As these are highly dependent on the implementing organisation, the BuNe presented provide an initial starting point, with the potential for further additions from case studies in industrial environments.

Nevertheless, this review shows a theoretical gap in the field of ECM. To aid organisations with managing design, production, supply chains, and EC, applicable tools for automatic handling of EC are necessary. Future research should thus focus on modelling the EC process and enabling their agent-based realisation. This can be done by suggesting optimal effectivity days, automating the EC scheduling task, and supervising the process. Regarding this, AI provides an opportunity. Additional exploration on what algorithms to use is required, and an agent-based approach to model and handle the complexity of ECM should be discussed. Finally, to confirm the theory, the resulting artefacts should be tested with industry partners.

3.9 Appendix

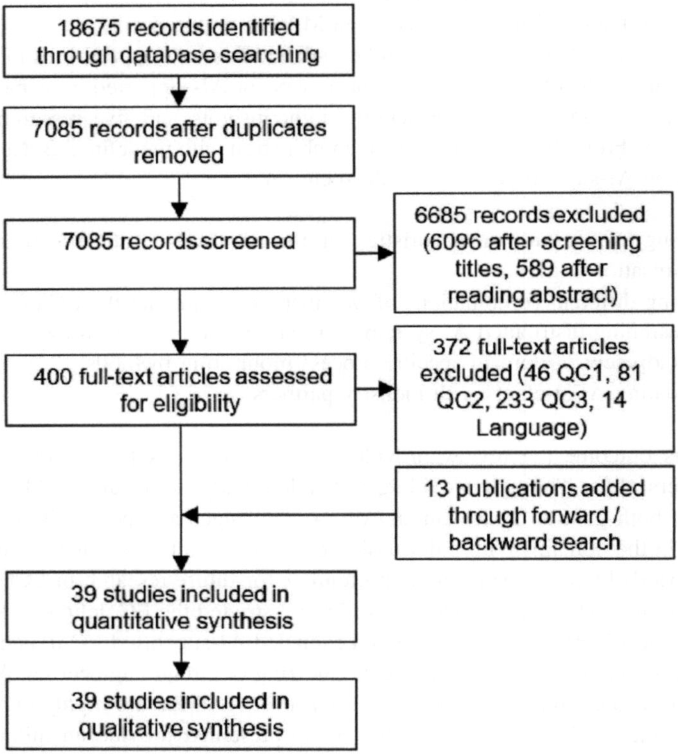

Fig. 3.12 PRISMA Diagram according to MOHER ET AL. [99]

3.10 Presentation of Own Contribution

All content presented within the systematic literature search was collected and processed by me based on the procedure model of VOM BROCKE ET AL. [82]. This includes definition of the research scope and research questions, identification of relevant data bases, definition of the search string, identification, definition and application of a suitable framework, analysis of literature by qualitative statistics and in-depth discussion, finalised by the subsequent synthesis and reflection. Furthermore, the textual elaboration and writing of the publication were entirely conducted by me. All four co-authors Prof. Dr. Burggräf, Dr. Wagner, Mr. Saßmannshausen and Dr. Weisser contributed with ideas to the research concept and in the internal review process.

Publication II: Deciding on When to Change—A Benchmark of Metaheuristic Algorithms for Timing Engineering Changes

4

4.1 Abstract

Changes to components, known as engineering changes (ECs), rarely occur on their own. In fact, in complex assembly systems, most ECs are introduced in batches to ensure that changed components match. As a result, to implement ECs optimally, multiple component's stock must be considered until the change is executed on the EC effectivity date. This problem is known as the EC effectivity date optimisation problem, a variation of the general inventory control problem with deterministic and dynamic demand. As optimisation and monitoring of this problem is computationally expensive, research has suggested to investigate whether metaheuristics can provide adequate support. To fill this research gap, we present the results of a benchmark on basic metaheuristics for EC effectivity date optimisation. To do so, we have compared five common metaheuristics in their basic form (Ant Colony Optimisation, Genetic Algorithm, Particle Swarm Optimisation, Tabu Search, and Simulated Annealing) on a real-world test set. Of the tested algorithms the Genetic Algorithm identified most best solutions and returned good average results for the test cases. However, as its reliability was comparatively low, our research suggests a sequential application of the Genetic Algorithm and Tabu Search.

Burggräf, P., Steinberg, F., Weißer, T., & Radisic-Aberger, O. (2024). Deciding on when to change—a benchmark of metaheuristic algorithms for timing engineering changes. International Journal of Production Research, 62(9), 3230–3250. https://doi.org/10.1080/00207543.2023.2226778

O. Radišić-Aberger, *Optimisation and Control of Engineering Change Schedules in the Automotive Industry with Metaheuristics and Machine Learning*, Findings from Production Management Research, https://doi.org/10.1007/978-3-658-51074-9_4

4.2 Introduction

Changes to components and products, though undesired, are necessary for manufacturing industries to remain profitable [40]. As consumer behaviour changes and products become more digitalised [15], these engineering changes (EC) occur more frequently. An EC, which is any modification to a component's structure, function, or behaviour [4], always results from a trigger [14], after which it is processed through a company-specific individualised process [30]. However, due to the phenomenon of change propagation [13], an EC on one component can cause more changes in other components. Especially in complex assembly systems, such as automotive and aerospace, a high variety of components and short product lifecycles lead to an increased number of ECs [14, 143]. Furthermore, supply chains and mass-customised products lead to frequent adaptation of production plans to meet customer demand [143]. To add difficulty to the problem, where an EC to one component causes ECs in others, both components must change simultaneously. To reduce administrative complexity of such interdependent changes, the combination of ECs into batches [63] or releases [15] has been devised. In this environment, determining and monitoring an optimal EC implementation plan becomes particularly challenging.

The resulting optimisation problem, generally known as EC effectivity date optimisation, is an inventory control problem. With these types of problems, the main challenge lies with inventory modelling. Literature knows three approaches to identify the best inventory control policy: Exact models, simulations, and (meta-)heuristics [25, pp. 501–530]. However, no exact model is applicable to the EC effectivity date optimisation problem in complex assembly systems. Moreover, as applying an exact method is computationally expensive, due to frequent updates of production and delivery figures the solution is potentially irrelevant by the time computation has finished. In these cases, TAHA [25, p. 530] suggests either using a simulation approach combined with a heuristic search, or a standalone heuristic search. As simulations are computationally expensive and are difficult to automate, it has been suggested to evaluate whether metaheuristic (MH) algorithms are applicable for EC effectivity date optimisation [1, 24].

Compared to exact methods like linear programming, MH algorithms compute only near optimal solutions [144, p. 23]. However, comparatively, they quickly find said solution even for combinatorial problems [25, p. 397] and are quick to adapt [20]. Although differing in their search strategy, they generally follow the same procedure: during each iteration step, new solutions are generated based on previous ones, evaluated against an objective via a cost-function, and if a better solution was discovered it is included for the next iteration cycle. Performance of MHs varies based on the application case, and no overall best has been identified [20, 145], further evidenced by several contributions within the EC domain [124, 125] and outside [146–148]. With efficiency and effectiveness being the main goals of EC management (ECM) research, similar comparisons, known as benchmarks, are necessary to evaluate MH application for EC. To our knowledge, this benchmark of MHs for the EC effectivity date optimisation problem has not been discussed in literature.

To address this research gap and set a base for future EC effectivity date research, this paper aims to evaluate which basic MH algorithm performs best for EC effectivity date optimisation in terms of (1) quality of solutions, (2) reliability, and (3) computational cost. To address this, five basic MH algorithms (Genetic Algorithm (GA), Ant Colony Optimisation (ACO), Particle Swarm Optimisation (PSO), Tabu Search (TS) and Simulated Annealing (SA)), were compared on a real-world test set of 50 EC batches from a German automotive OEM. For experiment preparation, we used the benchmarking approach by BEIRANVAND ET AL. [149]. Ultimately, our research targets at answering the research question: '*Can metaheuristic algorithms calculate EC effectivity dates sufficiently well?*'. By answering this question, this research provides three contributions:

- We present a novel model for EC effectivity date optimisation for usage with tabular production plan data.
- We show that, for the environment tested, the GA is significantly better than other base algorithms at finding the best EC effectivity date.
- We provide a blueprint on how to benchmark metaheuristics for EC effectivity date optimisation.

The remainder of our contribution is structured as follows: Sect. 4.3 provides an overview of existing methods and concepts for EC effectivity date optimisation. In Sect. 4.4, we define the used model as well as introduce the reader to the test set. Sect. 4.5 presents the results, whose implications are discussed in Sect. 4.6. In Sect. 4.7 a summary and outlook onto future research is given.

4.3 Theoretical Background

As stated, EC effectivity date optimisation is a variation of the general inventory control problem. In such problems, the general goal is to reduce inventory cost, while ensuring sufficient supply for assembly. Similarly, with EC effectivity date optimisation problems, the goal is to reduce the cost of obsolete inventory. However, as ECs are often triggered to reduce running cost [14], they become an additional decision criterion [23]. In complex assembly systems, material planners devise EC implementation plans such that the total cost is reduced to a minimum [17]. This is done through setting EC effectivity dates. They indicate days on which a component's state switches from a pre-EC state to a post-EC state. As multiple departments are tasked with ECM, the term effectivity date can indicate different points in time [55]. For logistics, it is the day on which the last time a pre-EC component is delivered. For production, it is the day on which the first post-EC component is consumed at the assembly line. To avoid confusion, in our research, we use the term EC effectivity date for indicating the change at the assembly line. The logistics' effectivity date is termed as *last call-off date*. The term *implementation scenario* describes a set of effectivity dates and last call-off dates for a given EC batch.

To reduce the cost of implementing ECs, research has devised quantitative and qualitative methods. In general, the methods apply the same procedure: Based on the trigger of the EC, a case-sensitive implementation scenario is drafted. However, the aforementioned batch strategy adds difficulty, as delivery plans are seldomly synchronised and workloads peak along project milestones [26]. As a consequence of running costs, the optimal implementation scenario is not necessarily to have every component's stock reduced to zero, but rather where the total sum of running and inventory costs is reduced to a minimum. To exemplify, Fig. 4.1 provides a showcase of three EC implementation scenarios. Without an inventory policy, any effectivity date would result in the total EC cost as shown in Fig. 4.1(a). However, with a smart inventory policy the stock levels can be modified pre-emptively, reducing those costs. Additionally, the optimal policy is not static. As the production figures change constantly, so does the optimal EC implementation scenario, necessitating constant monitoring.

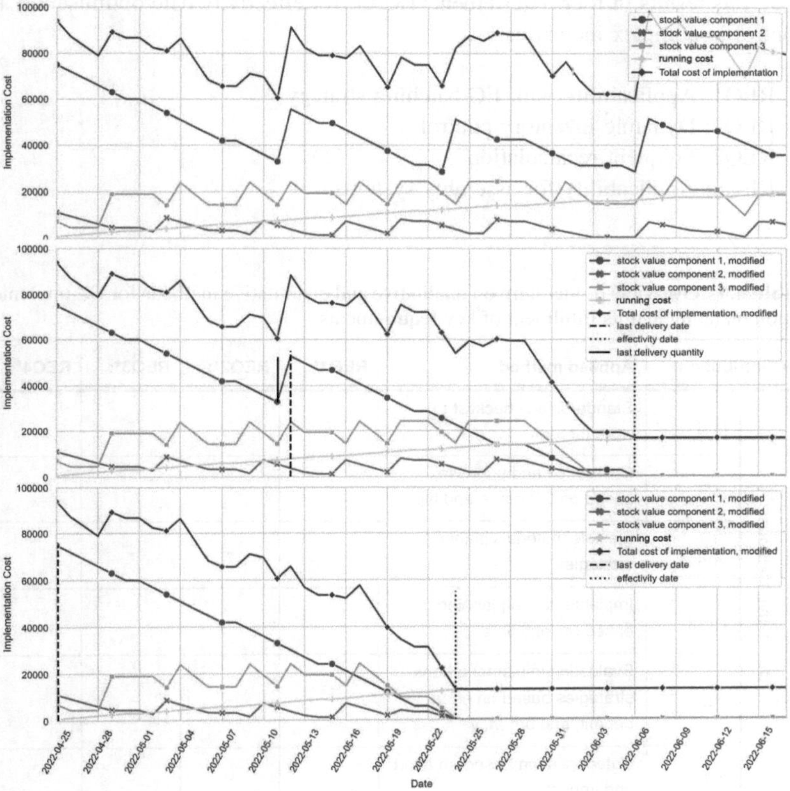

Fig. 4.1 The effects of different EC implementation scenarios. (**a**) shows the costs without any optimisation, (**b**) and (**c**) show possible alternatives (last call-off date and quantity only shown for component 1 for better readability)

To ensure efficiency and effectivity every EC's implementation scenario should be monitored closely. However, as at any time more than 1000 ECs can be active concurrently, advanced control systems are required [1]. By Ashby's law, such a control system would require at least the same degree of variety [19]. Foremost, to be applicable in complex assembly systems, a control system would have to be capable of determining optimal effectivity dates for entire EC batches at the same frequency as the production plan changes. Together with the automation goal, this results in four requirements (REQ) for effectivity date optimisation and control in complex industries:

- REQ1: Applicability with EC batching strategy
- REQ2: Dynamic inventory control
- REQ3: Frequent recalculation
- REQ4: Applicability for assembly systems

Table 4.1 Overview of literature on qualitative and quantitative methods for EC implementation planning and its fulfilment of key requirements

Contribution		Applied method	REQ1*:	REQ2*:	REQ3*:	REQ4*:
Qualitative Methods	[56]	Standardised checklist to work through	-	-	-	-
	[55]	Categorise and prioritise based on EC grade and type	-	-	-	-
	[26]	Flexible materials planning strategies	-	-	-	-
	[150]	Implementation planning based on lead time	-	-	-	-
	[15]	Evaluating implementations strategies based on effort, benefit, and urgency	-	-	-	-
	[27]	Categorisation based on effort and impact	-	-	-	-
Quantitative Methods	[23]	Linear programming model for specific categories			✓	✓
	[24]	Linear programming model on shop floor level	✓	✓		
	[7]	Simulation method based on date timing	✓	✓		✓
*Fulfilment of requirements as stated in Section 4.3						

*Fulfilment of requirements as stated in Sect. 4.3

Any optimisation method applied in such a control system would have to fulfil these four requirements. Table 4.1 shows a summary of the quantitative methods, as well as qualitative methods. Although all methods come to the same conclusion regarding the overall strategy, no quantitative method satisfies all four requirements. The linear programming method by BARZIZZA ET AL. [23] fails at calculating an optimal effectivity date for multiple components (REQ1). They simplify the problem by focusing solely on the most expensive component which would lead to increased scrapping costs for low-value goods. Additionally, they assume a static initial stock, rather than a deterministic dynamic replenishment (REQ2). The visual simulation method, similar to the example in Fig. 4.1(b) and Fig. 4.1(c) by CAPISTRANO BURGOS ET AL. [7] is able to provide solutions for multiple components, however for frequent recalculations (REQ3), the material planners would have to continuously monitor multiple graphs and manually test new combinations for delivery. Finally, the method introduced by SHIAU [24] incorporates multiple components, though their detailed model is targeted towards the shop floor, not assembly systems (REQ4). As such, there are no variables for externally sourced components or stock delivery reduction. Additionally, the exact calculation for each EC took approximately an hour, making frequent recalculations insensible (REQ3). As a result, they conclude that their model is not practically applicable, and they suggest investigating MHs. Thus, with no method being available, we investigate whether a heuristic optimisation method is sensible.

4.4 Material and Methods

4.4.1 Defining the Problem and Developing the Model

To test whether MHs can be satisfactorily applied for EC effectivity date optimisation in assembly systems and evaluate which base algorithm performs best a mathematical model was developed and five common basic MHs benchmarked. Section 4.4.1 introduces the model, while Sect. 4.4.2 discusses the benchmarking setup.

4.4.1.1 Model Assumptions

As introduced, the goal of EC effectivity date optimisation is to minimise the variable costs of implementing a specific EC batch. WÄNSTRÖM AND JONSSON [17] provide four actions to optimise the implementation scenario: Modifying the quantity delivered, the last call-off date, changing the EC effectivity date, and rescheduling the first call-off of the post-EC component. However, in complex assembly systems, to avoid line supply complexity, only three actions are relevant. As at no point in time pre-change and post-change variants of a component are allowed to be at the assembly line storage simultaneously, the first call-off of the new component is an offset of the effectivity date.

This results in three actions that directly describe the implementation scenario: Quantity reduction of the last delivery (1), changing the effectivity date (2), and changing the date of the last delivery (3) (Fig. 4.1(b)). To develop a model for testing MHs for EC effectivity date optimisation, we build on these actions and the existing models, although we modify some aspects based on the following assumptions (A1–A8):

- A1: The simultaneous use of pre-change and post-change components is prohibited to reduce line-site storage complexity.
- A2: To ensure parts availability and avoid stock outage at the assembly line, the earliest possible effectivity date is an offset of the earliest possible call-off date by the delivery lead time.
- A3: The first call-off date of a new component is an offset of the effectivity date as it reduces complexity at the assembly line. The offset is the delivery lead time.
- A4: Only ECs without a predetermined effectivity date are considered. In case of predetermination of effectivity dates due to the EC change type (e.g., mandatory EC), the decision variables would be reduced to two, which would simplify the model.
- A5: Costs are summarised as running costs (e.g., warranty, rework), inventory costs (i.e., scrapping costs), and penalty costs (e.g., costs of changing call-offs). Costs for tool change, regular transport, EC administration etc. are not assessed.
- A6: When multiple ECs are batched into a single EC batch, either all components must be introduced simultaneously (type II), or none must (type I).
- A7: It is assumed that the implementation optimisation takes place before the delivery plan is frozen. Before the freeze of the delivery plan, no economic

downsides result from changing the effectivity date. At our industry partner this would be 21 days before[1].

- A8: Logistical data input is expected to come directly as MRP [151] or MRP-II [152] excerpts, and is transformed into pairs of arrays.

4.4.1.2 Nomenclature

With the given notation, the previously described problem definition, objective, and assumption can be translated to a mathematical model. For readability, all sets are indicated by curly uppercase letters and corresponding indices written as subscripts. Variables representing a date are indicated by a lowercase d, with the superscript letter indicating the specifics.

Indices and sets:

- $\mathcal{J}\ominus := \{1, \ldots, j, \ldots J\}$, the set of components affected by an EC batch
- $\mathcal{F} := \{1, \ldots, f, \ldots F\}$, the set of factories affected by an EC batch
- $\mathcal{D} := \{1, \ldots, d, \ldots D\}$, set of possible dates for the effectivity date
- $\mathcal{J}_f$, the subset of components being consumed at factory f

Parameters:

- rc_{jf}: Sum of running costs associated with every pre-change component j at factory f (e.g., rework, warranty)
- ic_{jf}: Sum of inventory costs associated with component j which is not used for production (mainly scrapping costs) in factory f
- pc^b_{jf}: : Sum of base penalty costs occurring due to additional handling costs (e.g., changing delivery plans) for component j at factory f
- pc^h_{jf}: : Sum of penalty costs occurring due to additional handling (e.g., changing delivery plans) for component j at factory f
- DL_{jf}: Delivery lead time of component j to factory f

Decision variables:

- y^l_{jf}: Delivery quantity of last delivery for component j at factory f in percent
- d^e_{jf}: Effectivity date of EC for component j at factory f
- d^l_{jf}: : Phase-out date (last delivery date) for component j at factory f

---

[1] Within this 21-day window, although changes to deliveries are still possible, they are restrictively used. As they result in additional complexity and additional danger of stock outage, they are only accepted to advert immediate danger to the assembly process

Input variables:

- d^0_{jf}: The earliest possible call-off date for component j at factory f
- s^{pre}_{0jf}: Current stock of component j at factory f
- X^{pre}_{jf}: Volume consumed of component j consumed at factory f until the effectivity date d^e_{jf}
- X^{del}_{jf}: Volume delivered of component j delivered to factory f until the effectivity date d^e_{jf}

4.4.1.3 Model Presentation

Each EC batch consists of an $\mathcal{N}$ number of ECs. We are given sets of $\mathcal{J}$ components and $\mathcal{F}$ factories affected by a given EC batch. As some components are not necessarily consumed at each factory, there can be less ECs in an EC batch than combinations of j and f, but never more. Thus, $\mathcal{N} \leq \mathcal{J} \times \mathcal{F}$. Let $\mathcal{J}_f$ indicate the subset of $j \in \mathcal{J}$ components which are consumed at factory f. Each EC $n \in \mathcal{N}$ describes the transition of component $j \in \mathcal{J}_f$ at factory $f \in \mathcal{F}$ from a pre-change component state *pre* to a post-change component state *post*. Let $Cost^{Batch}$ indicate the cost resulting from implementing the full EC batch into production, the resulting total cost is the sum of $Cost^{EC}_n$ for every individual EC $n \in \mathcal{N}$ (Equation (4.1)). Equation (4.1) can in turn be rewritten as the summated EC cost for component $j \in \mathcal{J}_f$ at a given factory $\in \mathcal{F}$ $Cost^{EC}_{jf}$, summed over all affected factories $f \in \mathcal{F}$ (Equation (4.2)).

$$Cost^{Batch} = \sum_{n \in N} Cost^{EC}_n \tag{4.1}$$

$$\sum_{n \in \mathcal{N}} Cost^{EC}_n = \sum_{f \in \mathcal{F}} \sum_{j \in \mathcal{J}_f} Cost^{EC}_{jf} \tag{4.2}$$

The variable cost of each EC $\left(Cost^{EC}_{jf}\right)$ from changing component j in factory f from a state *pre* to *post*, is defined as the total sum of running costs $\left(RC_{jf}\right)$, obsolete inventory costs $\left(IC_{jf}\right)$, and total penalty costs $\left(PC_{jf}\right)$. Thus $Cost^{EC}_{jf}$, can be written as

$$Cost^{EC}_{jf} = RC_{jf} + IC_{jf} + PC_{jf} \tag{4.3}$$

specifically:

RC_{jf} (the total running cost of component j in factory f) is the product of variable costs rc_{jf} multiplied by volume consumed X_{jf}^{pre}, which in turn is depends on the effectivity date d_{jf}^{e} (Equation (4.4)).

$$RC_{jf} = rc_{jf} \times X_{jf}^{pre}\left(d_{jf}^{e}\right) \tag{4.4}$$

IC_{jf} (total inventory costs of component j in factory f) are in essence scrapping costs. They are costs resulting from scrapping leftovers of component j in factory f after the switch from *pre* to *post* and are the product of the unit cost sc_{jf} and the volume on stock on the effectivity date $s_{jf}^{pre}\left(d_{jf}^{e}, d_{jf}^{l}\right)$ (Equation (4.5)).

$$IC_{jf} = ic_{jf} \times s_{jf}^{pre}\left(d_{jf}^{e}, d_{jf}^{l}\right) \tag{4.5}$$

As indicated by the second variable, the stock function $s_{jf}^{pre}\left(d_{jf}^{e}, d_{jf}^{l}\right)$ is not only dependent on the effectivity date d_{jf}^{e}, but also on the date of the last call-off d_{jf}^{l}.

Furthermore, it is not a linear function. It is, as shown in Fig. 4.1, a reverse sawtooth curve with varying frequency and amplitude. The amplitude of each tooth represents stock replenishment. The resulting stock on the effectivity date d_{jf}^{e} can be calculated by Equation (4.6). It is the sum of the initial stock s_{0jf}^{pre} and the cumulative sum of all deliveries of component j $\left(X_{jf}^{del}\left(d_{jf}^{l}\right)\right)$ until the last call-off date $\left(d_{jf}^{l}\right)$, reduced by the cumulative volume consumed $X_{jf}^{pre}\left(d_{jf}^{e}\right)$.

$$s_{jf}^{pre}\left(d_{jf}^{e}, d_{jf}^{l}\right) = s_{0jf}^{pre} + X_{jf}^{del}\left(d_{jf}^{l}\right) - X_{jf}^{pre}\left(d_{jf}^{e}\right) \tag{4.6}$$

PC_{jf} (total penalty costs), are costs which occur due to exceptional activities. This could be for instance the cost of speeding up a delivery, a penalty for switching the effectivity date after the delivery freeze window (cf. A7), or similar. In our case, as we want to test MHs rather than develop a detailed model, we only include costs of changing the quantity delivered with the last delivery (Equation (4.7)). This can be interpreted as reducing the amplitude of the tooth indicated by the last call-off date d_{jf}^{l} (i.e., last delivery) of the sawtooth wave in Fig. 4.1(b).

Other costs could be similarly included, although some might add more decision variables. Following consultation by the industry partner, we have modelled the cost as a linear function, with a base cost pc_{jf}^{b}, and a variable component based on the proportional reduction y_{jf}^{l} and a factor pc_{jf}^{h}.

$$PC_{jf} = pc_{jf}^{b} + pc_{jf}^{h} \times \left(1 - \frac{y_{jf}^{l}}{100}\right) \tag{4.7}$$

The goal is to minimise the overall cost of implementing the entire EC batch. With the given Equations (4.[4.1–4.7]) the optimisation problem can be stated as:
Find$d_{jf}^{e} \in \mathcal{D}$, $d_{jf}^{l} \in \mathcal{D}$,$y_{jf}^{l} \in 0, 100$, which:

Objective function:

$$\begin{aligned} Minimise \sum_{f \in F} \sum_{j \in f} rc_{if} \times X_{if}^{pre}\left(d_{jf}^{e}\right) + ic_{jf} \times \left(s_{0}{}_{if}^{pre} + X_{jf}^{del}\left(d_{jf}^{l}\right) - X_{jf}^{pre}\left(d_{jf}^{e}\right)\right) \\ + pc_{jf}^{b} + pc_{jf}^{h} \times \left(1 - \frac{y_{jf}^{l}}{100}\right) \end{aligned} \tag{4.8}$$

subject to constraints:

$$s_{0}{}_{if}^{pre} + X_{jf}^{del}\left(d_{jf}^{l}\right) - X_{jf}^{pre}\left(d_{jf}^{e}\right) \geq 0 \qquad \forall j \in \mathcal{J}_f; \forall f \in F \tag{4.9}$$

$$d_{jf}^{e} = \begin{cases} d_{1f}^{e} = d_{2f}^{e} \ldots = d_{jf}^{e} \text{ if parts are interdependent} \\ d_{1f}^{e},\ d_{2f}^{e},\ \ldots d_{jf}^{e} \text{ else} \end{cases} \qquad \forall j \in \mathcal{J}_f; \forall f \in F \tag{4.10}$$

$$d_{jf}^{e} \geq d_{jf}^{0} + DL_{jf} \qquad \forall j \in \mathcal{J}_f; \forall f \in F \tag{4.11}$$

Constraint (4.9) ensures that the stock never becomes negative. Constraint (4.10) controls whether components must be introduced simultaneously (type II), or whether independent implementation is possible (type I) (cf. A6). Constraint (4.11) controls that the effectivity date d_{jf}^{e} cannot be earlier than the earliest call-off d_{jf}^{0}, offset by the delivery lead time DL_{jf} (cf. A2).

4.4.2 Benchmarking of Algorithms

Benchmarking is the systematic comparison of multiple products, in our case metaheuristics, with a given set of metrics [149]. As a guiding methodology, we apply the benchmarking procedure of optimisation algorithms as shown by BEIRANVAND ET AL. [149].

4.4.2.1 Reason for Benchmarking Metaheuristics for EC Effectivity Date Optimisation

As shown throughout the paper, EC effectivity date optimisation is a complex real-world problem. It is not sufficient to determine the implementation scenario once, but every time the production and delivery plans are updated, which lead to the suggestion to test MH algorithms. At our industry partner, the update occurs every 15 minutes. However, no overall best MH has been identified and an application context specific investigation is necessary [145]. Thus, we perform a benchmark of MH for EC effectivity date optimisation. From the use-case description, we define the relevant metrics (1) quality of solution, (2) reliability, and (3) computational cost.

4.4.2.2 Description of Test Sets

As we want to understand which basic MH performs best for the real-world problem of EC effectivity date optimisation, the test set should accordingly be from the real world [149]. To provide reliable results, 50 EC batches are randomly chosen from all currently pending EC batches at the industry partner. In Fig. 4.2, the test set is graphically shown along the axes of 'number of components' and 'number of factories'. The distribution of test cases matches the overall distribution of all EC batches at the industry partner. From a computational point of view, only the number of ECs in an EC batch determines the complexity. From an organisational perspective, a globally complex EC batch introduces additional difficulties, as delivery lead times might vary drastically based on factory location and involve more personnel.

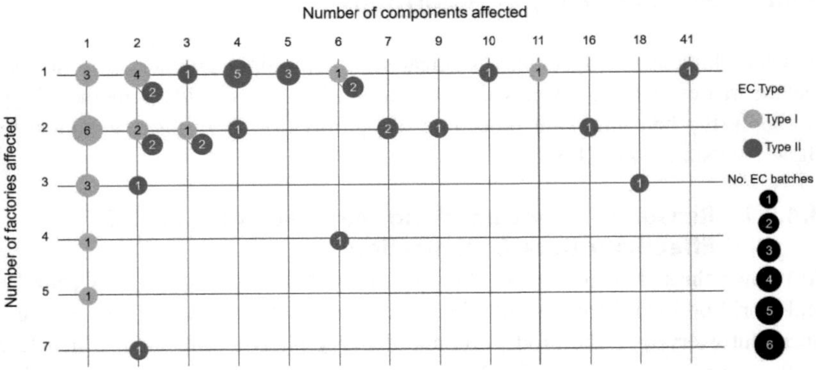

Fig. 4.2 Complexity distribution of test cases. Where types overlap, the number of hidden cases is shown. Detailed tabular description can be found in the Appendix, Table 4.5

4.4.2.3 Solution Encoding and Search Space

As seen from the mathematical model, three variables define an implementation scenario. Hence, the solution of the algorithms for each EC can be represented by a triplet of numbers (z_1, z_2, z_3). In our model, all three numbers represent integer values, where z_1 corresponds to y^l, the relative percentage of goods delivered. Correspondingly, z_2 and z_3 represent the effectivity date d^e and the last call-off date d^l. The range was defined as 0 to 100, as the industry partner provides detailed daily production plans for approximately 100 days after the delivery plan freeze (cf. A7, sec. 4.4.1.1). Zero was included, as in theory, an immediate stop could also be beneficial. The values z_2 and z_3 represent the position of a date chosen in a possible date array of length 101 and are matched with the production and delivery plan data received from the MRP-II system. To improve the heuristic search, the earliest possible date is calculated before the optimisation as the one which satisfies Equation (4.9).

To give an example: For the case and implementation scenarios presented in Fig. 4.1, Fig. 4.1(b) would be the implementation scenario encoded as (89, 43, 17; 95, 43, 33; 99, 43, 33). It would result in a change on the 43rd possible day, reducing the last delivery of component 1 on the 17th day to 89%. Similarly, deliveries of components 2 and 3 on the 33rd day would be reduced to 95% and 99% respectively. This would result in the EC being implemented on June 6th, with a total cost of approx. 17000$. For the same case, the solution (100, 30, 0; 50, 30, 25; 100, 30, 25) would result in approx. 13000$ (Fig. 4.1(c)), saving

4000$ due to reduced consumption costs and lower scrapping costs of component 2 and 3. All algorithms use the same solution encoding method.

4.4.2.4 Tested Algorithms

For comparison, five MH algorithms were chosen, which are already commonly used. These include the two trajectory search algorithms Tabu Search (TS) [153], and Simulated Annealing (SA) [154] and the three population-based algorithms Genetic Algorithm (GA) [155, 156], Ant Colony Optimisation (ACO) [157], and Particle Swarm Optimisation (PSO) [158]. Those five algorithms were chosen, as they are already applied within the ECM domain (e.g., GA and/or PSO in [20, 124, 125]) or have benchmarks in other domains which enables comparability (e.g., [146, 147, 159]. The algorithms were coded in Python programming language following the GA template given by KALAMI HERIS [160] and modified for discrete problems according to the descriptions presented in [25, 144, 146][2]. As no exact method and model suit the problem of EC effectivity date optimisation for multiple components in complex assembly systems (cf. sec. 4.3), no comparison to existing methods in literature can be given.

For testing, the algorithms were adapted to fit the EC domain and include heuristic information if possible. First, as EC implementation is a discrete optimisation problem, all algorithms were adapted to only present integers as a possible solution. Where randomisation and bounds are necessary, the algorithms share the same function, e.g., all random choices are incorporated by roulette wheel selection. Additionally, only feasible solutions were included in the set of solutions to derive new children (GA), reinforce the pheromone matrix (ACO), replace the best swarm solution (PSO), or set starting position for subsequent iterations (TS and SA).

In the next step, each algorithm was adapted to increase its chances of discovering a better solution. For ACO, the pheromone matrix was used to ensure that only call-off dates before the effectivity date could be chosen (as a later delivery is always fully scrapped and thus insensible). The length of the TS tabu-list was programmed as a product of the number of factories and parts to dynamically adapt in length. The GA was implemented as an intermediate cross-over algorithm with normally distributed mutation, where all genes of the parents were crossed to generate the offspring, and a random set of genes mutated.

[2] The depiction of the pseudo code is presented in sect. 4.8 Appendix, whereas the actual code used can be viewed in the online repository linked in the data availability statement.

The last step before running the experiments is parameter tuning. To set a base, we tuned the initial parameters by random search so that every algorithm used was able to discover a solution with 95% reliability for the example shown in Fig. 4.1. The random search has shown that for the GA, PSO, TS, and SA a relatively wide search space is required to find solutions reliably. Thus, for the GA, TS, and SA the σ-value, used to generate discover new solutions from currently available ones, was set to ± 15. For the PSO, implemented as a Geometric PSO [144], the velocity V_{max} was limited to ± 15. From there, every algorithm had its parameters defined, and no tuning was done as the problems' complexity varied [149]. The parameters are summarised in Table 4.2.

Table 4.2 Parameter set for the algorithms

Parameter*	ACO	GA	PSO	TS	SA
Iterations	1000	1000	1000	1000	1000
Population	50	50	50	20 (size of neighbourhood)	–
Parent / Child Ratio	–	1	–	–	–
Search space (σ)	–	± 15	–	± 15	± 15
γ	–	0,9 (crossover probability)	–	–	–
μ	–	0,5 (mutation rate)	–	–	–
α	1 (pheromone influence)	–	–	–	0,99 (cooling factor)
β	0 (heuristic influence)	–	–	–	–
ρ	0,005 (evaporation rate)	–	–	–	–

(continued)

Table 4.2 (continued)

Parameter*	ACO	GA	PSO	TS	SA
ρ1	–	–	0,8 (Acceleration coefficient 1)	–	–
ρ2	–	–	0,3 (Acceleration coefficient 2)	–	–
Vmax	–	–	±15 (maximum velocity)	–	–
Temperature	–	–	–	–	2000
Temperature holding duration	–	–	–	–	10

*letter notation as shown in the publications introducing the algorithms [153–158]

4.4.2.5 Running of Experiments

To have an equal testing ground for all algorithms and evaluate the criteria as defined in 4.4.2.1, the starting point of each implementation scenario optimisation was chosen randomly in advance and shared across all algorithms for each run. For non-population-based algorithms (TS and SA), the best initial solution was taken as the starting point. For enabling comparability to [146] where the same algorithms were tested, we have evaluated each test case 200 times for each algorithm, with randomly varying starting points. The full computations were run in one go in sequence on an AWS Sagemaker (ml.g4dn.xlarge) instance.

Regarding the quality of the solution, additional considerations are necessary. As the optimal solution is not known, only a fixed-cost approach on running the algorithms can be applied [149]. As there exists no clear identification on how many iterations should be run, we set the number of iterations to 1000. We chose this value as to limit computational resource consumption, as during parameter optimisation on the test case, the slowest algorithm was already taking approximately 1 minute per run of 1000 iterations. Accordingly, the best solution of any algorithm was taken as the overall optimal solution [149]. Furthermore, as we run the algorithms multiple times, the mean solution cost is an indicator on the robustness of the algorithm. As calculations are run every time a new production plan becomes available, it can be more beneficial to have robust mediocre solutions, instead of widespread solutions. Thus, we also evaluate the mean solution value. For investigating whether the outcome of multiple algorithms is statistically significant, DEMŠAR [161] suggest a Friedman test [162] followed by a post-hoc Nemenyi test [163].

Reliability (2), determined as the second criterion, is calculated as the identification frequency of a solution within a 5% range of the best possible solution (or mean solution) vs. the total number of experiments per algorithm [149]. As a combination of (2) and (3), we also show the success rate vs. average iterations to find a solution within 5% of the best (or mean) solution as an indicator for performance.

For (3) we have chosen the wall-clock time, as suggested in [149]. As this metric is highly hardware dependent, we additionally provide a relative comparison. The calculation is done by timestamps at the beginning and end of each run, and the average difference is shown.

4.5 Results

4.5.1 Quality of Solutions: Best Cost and Mean Cost

For each of the 50 test cases, the optimal effectivity date, last call-off date and call-off quantity and its associated costs were calculated. Table 4.3 presents the best solution cost of any algorithm whereas Table 4.4 shows the mean solution cost. Furthermore, each algorithm's best implementation cost is shown as a factor of the best cost found by any algorithm. Fig. 4.3 (a) and (b) show the Nemenyi test corresponding to the ranked solution quality. Fig. 4.4 shows a visual representation of the best and mean solution quality for each EC batch.

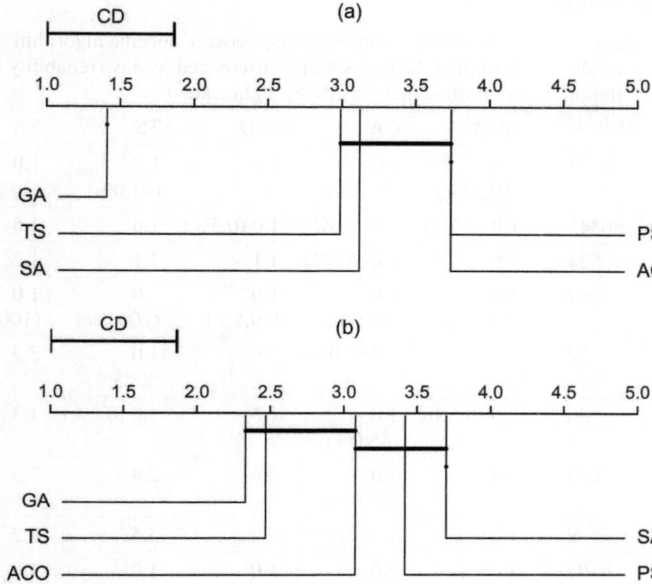

Fig. 4.3 Comparison of all metaheuristics against each other with the Nemenyi test in term of best (a) and mean (b) solution found. Groups of classifiers that are not significantly different (at $p < 0.05$) are connected

Table 4.3 Overview of best implementation costs and the comparative performance

EC batch ID	Best overall solution value [$]	Factored best implementation cost of specific algorithm regarding the best solution discovered by any (reliability to find solution within 5% of best in brackets)				
		ACO	GA	PSO	TS	SA
EC 1	290	1.0 (98.5%)	1.0 (38.0%)	1.0 (21.5%)	1.0 (100.0%)	1.0 (99.0%)
EC 2	14	1.0 (37.5%)	1.0 (38.5%)	2.8	1.0 (88.0%)	1.0 (42.0%)
EC 3	14	1.0 (25.5%)	1.0 (17.5%)	1.0 (55.0%)	1.0 (100.0%)	1.0 (100.0%)

(continued)

Table 4.3 (continued)

EC batch ID	Best overall solution value [$]	Factored best implementation cost of specific algorithm regarding the best solution discovered by any (reliability to find solution within 5% of best in brackets)				
		ACO	GA	PSO	TS	SA
EC 4	1773	1.0 (100.0%)	1.0 (27.5%)	1.1	1.0 (41.0%)	1.0 (30.0%)
EC 5	56345	1.0 (4.5%)	1.0 (5.0%)	1.0 (0.5%)	1.6	1.6
EC 6	524	1.5	1.0 (5.0%)	1.1	1.1	1.2
EC 7	684	1.0 (95.5%)	1.0 (10.0%)	1.0 (99.5%)	1.0 (100.0%)	1.0 (100.0%)
EC 8	22	7.8	1.0 (3.0%)	3.6	1.0 (28.5%)	3.5
EC 9	56	1.0 (0.5%)	1.0 (45.0%)	1.5	1.0 (0.5%)	1.3
EC 10	623	3.6	1.0 (34.0%)	3.0	2.4	2.3
EC 11	9	13.4	1.0 (5.0%)	8.2	3.5	8.2
EC 12	6197	1.0 (100.0%)	1.0 (100.0%)	1.0 (99.5%)	1.0 (100.0%)	1.0 (100.0%)
EC 13	11	15.5	1.0 (7.0%)	2.8	1.0 (0.5%)	2.8
EC 14	973	1.1	1.0 (10.5%)	1.0 (96.0%)	1.0 (17.5%)	1.0 (21.5%)
EC 15	1889	1.9	1.0 (10.5%)	6.6	1.0 (0.5%)	2.1
EC 16	3256	3.4	1.0 (10.0%)	3.5	1.4	1.2
EC 17	20993	13.8	1.0 (1.0%)	13.8	11.8	9.3
EC 18	15139	1.0 (100.0%)	1.0 (100.0%)	1.0 (100.0%)	1.0 (100.0%)	1.0 (100.0%)
EC 19	100	1.5	1.0 (37.5%)	4.5	1.0 (1.0%)	3.0
EC 20	30	14.3	1.0 (2.5%)	27.7	12.3	10.4
EC 21	336	2.6	1.0 (7.0%)	5.3	3.2	2.8
EC 22	310844	1.0 (100.0%)	1.0 (100.0%)	1.0 (100.0%)	1.0 (100.0%)	1.0 (100.0%)

(continued)

Table 4.3 (continued)

EC batch ID	Best overall solution value [$]	Factored best implementation cost of specific algorithm regarding the best solution discovered by any (reliability to find solution within 5% of best in brackets)				
		ACO	GA	PSO	TS	SA
EC 23	8027	1.2	1.0 (10.0%)	3.9	1.1	1.4
EC 24	6073	1.0 (100.0%)	1.0 (100.0%)	1.0 (100.0%)	1.0 (100.0%)	1.0 (100.0%)
EC 25	114	3.1	1.0 (18.0%)	5.2	2.0	1.7
EC 26	317	1.4	1.0 (69.0%)	1.8	1.0 (12.0%)	1.1
EC 27	67332	1.8	1.4	1.0 (98.0%)	1.8	2.2
EC 28	624	13.8	1.0 (0.5%)	15.1	16.5	16.3
EC 29	433	3.9	1.0 (9.5%)	2.9	2.8	6.5
EC 30	40388	1.0 (56.0%)	1.0 (34.5%)	2.0	1.0 (80.5%)	1.0 (43.5%)
EC 31	34647	1.0 (100.0%)	1.0 (84.5%)	1.0 (90.5%)	1.0 (100.0%)	1.0 (96.0%)
EC 32	661026	1.1	1.0 (100.0%)	1.0 (0.5%)	1.1	1.1
EC 33	1854	3.5	1.0 (1.0%)	2.9	3.7	3.8
EC 34	600	1.2	1.0 (18.5%)	1.5	1.1	1.1
EC 35	3222	2.6	1.0 (70.0%)	1.2	1.1	1.1
EC 36	0	Inf	1.0 (4.5%)	1.0 (72.0%)	Inf	Inf
EC 37	137	4.8	1.0 (1.5%)	3.4	3.9	4.7

(continued)

Table 4.3 (continued)

EC batch ID	Best overall solution value [$]	Factored best implementation cost of specific algorithm regarding the best solution discovered by any (reliability to find solution within 5% of best in brackets)				
		ACO	GA	PSO	TS	SA
EC 38	525285	5.4	1.0 (1.0%)	1.1	4.8	5.3
EC 39	137	3.8	1.0 (33.5%)	3.9	2.2	4.2
EC 40	68501	1.0 (38.0%)	1.0 (86.5%)	1.1	1.0 (7.5%)	1.0 (10.5%)
EC 41	42668	3.8	1.0 (10.5%)	2.1	2.4	1.7
EC 42	900	1.3	1.0 (72.0%)	1.5	1.3	1.2
EC 43	5419	1.4	1.2	1.0 (0.5%)	1.7	1.6
EC 44	167	5.7	1.0 (15.0%)	1.3	2.0	4.2
EC 45	13173	2.9	1.0 (0.5%)	1.4	3.3	2.7
EC 46	1100	1.5	1.0 (94.5%)	1.8	1.5	1.5
EC 47	2100923	1.0 (100.0%)	1.0 (98.0%)	1.0 (98.0%)	1.0 (86.0%)	1.0 (65.0%)
EC 48	10873	1.1	1.0 (98.5%)	1.1	1.5	1.1
EC 49	2849	1.4	1.0 (30.0%)	1.5	1.9	1.4
EC 50	25195	1.3	1.2	1.0 (2.0%)	2.5	2.3
Success rate (Av.reliability):		15/50 (70%)	47/50 (37%)	16/50 (65%)	20/50 (58%)	14/50 (72%)

Table 4.4 Overview of mean implementation costs and the comparative performance

EC batch ID	Best overall mean solution value [$]	Factored mean implementation cost of specific algorithm regarding the best mean solution discovered by any (reliability to find solution within 5% or better of mean in brackets)				
		ACO	GA	PSO	TS	SA
EC 1	962	2.7	1.0 (80.5%)	2.6	4.0	5.0
EC 2	42473	1.0 (90.5%)	4.1 (34.5%)	2.0	1.0 (83.5%)	3.0 (45.0%)
EC 3	68	6.1	2.4 (40.5%)	3.6 (2.5%)	1.0 (89.5%)	4.3 (14.5%)
EC 4	196583	111.4	1.6 (41.0%)	1.0 (72.0%)	4.2	13.9
EC 5	11574	1.0 (100.0%)	1.0 (98.5%)	1.0 (100.0%)	2.3	1.5 (0.5%)
EC 6	6201	1.0 (100.0%)	1.0 (100.0%)	1.0 (99.5%)	1.0 (100.0%)	1.0 (100.0%)
EC 7	29	8.9	1.0 (85.0%)	10.9	14.1	18.4
EC 8	4382	4.3 (0.5%)	1.4 (72.0%)	1.7 (65.5%)	1.0 (72.0%)	4.2 (24.0%)
EC 9	104	2.7	1.2 (56.0%)	3.7 (2.0%)	1.0 (42.0%)	2.4 (1.0%)
EC 10	16	2.1 (37.5%)	5.3 (38.5%)	13.0	1.0 (88.0%)	3.1 (42.0%)
EC 11	13146	1.3 (9.5%)	1.0 (71.0%)	2.8 (2.5%)	1.6 (6.0%)	1.6 (4.0%)
EC 12	495	3.8 (0.5%)	1.0 (55.5%)	4.2	2.6 (0.5%)	2.9 (0.5%)
EC 13	30463	1.3	1.0 (37.0%)	1.0 (50.5%)	2.0	2.0

(continued)

Table 4.4 (continued)

EC batch ID	Best overall mean solution value [$]	Factored mean implementation cost of specific algorithm regarding the best mean solution discovered by any (reliability to find solution within 5% or better of mean in brackets)				
		ACO	GA	PSO	TS	SA
EC 14	2320	1.5 (3.0%)	1.0 (48.5%)	1.5 (15.0%)	1.3 (29.0%)	1.2 (33.0%)
EC 15	105213	5.3	1.0 (22.0%)	1.9 (1.0%)	1.6 (1.5%)	1.5 (5.0%)
EC 16	14	1.7 (25.5%)	3.9 (17.5%)	9.3 (55.0%)	1.0 (100.0%)	1.0 (100.0%)
EC 17	34697	1.1 (8.0%)	1.1 (1.5%)	1.0 (64.5%)	2.2	2.1
EC 18	7993	1.0 (96.5%)	1.1 (69.5%)	1.2 (58.5%)	2.4	3.1
EC 19	6091	1.0 (100.0%)	1.0 (100.0%)	1.0 (100.0%)	1.0 (100.0%)	1.0 (100.0%)
EC 20	68351	2.1	1.7	1.0 (98.0%)	2.8	3.1
EC 21	289	1.0 (98.5%)	2.1 (38.0%)	1.2 (21.5%)	1.0 (100.0%)	1.0 (99.0%)
EC 22	13341	3.8 (1.0%)	1.0 (76.0%)	6.5	3.2 (2.0%)	5.9 (0.5%)
EC 23	661032	1.2	1.0 (100.0%)	1.1 (0.5%)	1.1	1.1
EC 24	684	1.0 (95.5%)	1.3 (10.0%)	1.0 (99.5%)	1.0 (100.0%)	1.0 (100.0%)
EC 25	9781	1.7	1.0 (37.5%)	2.9	1.5 (2.5%)	1.6 (5.5%)
EC 26	1817	1.0 (100.0%)	2.9 (27.5%)	1.1 (96.0%)	1.6 (41.0%)	1.9 (30.0%)
EC 27	310843	1.0 (100.0%)	1.0 (100.0%)	1.0 (100.0%)	1.0 (100.0%)	1.0 (100.0%)

(continued)

Table 4.4 (continued)

EC batch ID	Best overall mean solution value [$]	Factored mean implementation cost of specific algorithm regarding the best mean solution discovered by any (reliability to find solution within 5% or better of mean in brackets)				
		ACO	GA	PSO	TS	SA
EC 28	2101598	1.0 (100.0%)	1.0 (98.0%)	1.0 (98.0%)	1.0 (86.0%)	1.2 (65.0%)
EC 29	191	3.9	1.0 (88.0%)	2.9	2.7	4.1
EC 30	3609	2.3	1.0 (88.0%)	1.4	1.1 (25.5%)	1.3 (5.0%)
EC 31	425	1.8	1.0 (77.0%)	1.3	1.7	2.1
EC 32	90667	1.0 (100.0%)	1.2 (88.0%)	1.0 (95.5%)	1.0 (99.5%)	1.1 (95.0%)
EC 33	1244	1.0 (100.0%)	5.2 (88.0%)	11.5	1.1 (84.0%)	16.5 (92.5%)
EC 34	296	1.2 (6.5%)	12.8 (38.5%)	11.3	1.0 (92.0%)	41.8 (32.0%)
EC 35	335	1.4 (0.5%)	2.7 (46.0%)	3.4	1.0 (65.0%)	1.8 (32.0%)
EC 36	910	1.2 (24.5%)	1.0 (84.5%)	1.3 (2.5%)	1.1 (42.5%)	1.2 (27.0%)
EC 37	6186	1.1	1.0 (22.5%)	1.0 (91.0%)	1.1	1.5
EC 38	15139	1.0 (100.0%)	1.0 (100.0%)	1.0 (100.0%)	1.0 (100.0%)	1.0 (100.0%)
EC 39	213	5.1	1.0 (99.5%)	1.0 (100.0%)	2.4	4.5
EC 40	64	1.4 (0.5%)	1.0 (45.0%)	1.3	1.1 (0.5%)	1.3
EC 41	74406	1.0 (100.0%)	1.4 (88.5%)	2.4 (91.5%)	1.0 (100.0%)	3.2 (90.0%)

(continued)

Table 4.4 (continued)

EC batch ID	Best overall mean solution value [$]	Factored mean implementation cost of specific algorithm regarding the best mean solution discovered by any (reliability to find solution within 5% or better of mean in brackets)				
		ACO	GA	PSO	TS	SA
EC 42	354	1.4	15.0 (71.5%)	2.4	1.0 (99.0%)	11.0 (5.5%)
EC 43	132590	6.6	1.0 (69.0%)	2.5	5.5	6.5
EC 44	744089	12.7	1.6 (13.5%)	1.0 (87.5%)	6.0	8.6
EC 45	3002	1.3	1.0 (100.0%)	1.4	2.9	1.9
EC 46	1125	1.6	1.0 (95.5%)	3.2	1.6	1.6
EC 47	34781	1.0 (100.0%)	1.1 (84.5%)	1.0 (90.5%)	1.0 (100.0%)	1.0 (96.5%)
EC 48	993	1.2 (1.0%)	2.0 (10.5%)	1.0 (96.0%)	1.2 (17.5%)	1.2 (21.5%)
EC 49	6911	1.9 (1.5%)	1.0 (47.5%)	5.4	2.0 (1.5%)	2.6 (2.5%)
EC 50	795	1.0 (100.0%)	8.8 (23.0%)	3.3	5.3 (56.5%)	8.6 (26.5%)
Success rate (Av. reliability)		30/50 (57%)	49/50 (62%)	30/50 (65%)	33/50 (61%)	33/50 (45%)
Av. relative standard deviation		0.25%	1.77%	0.81%	0.38%	5.67%

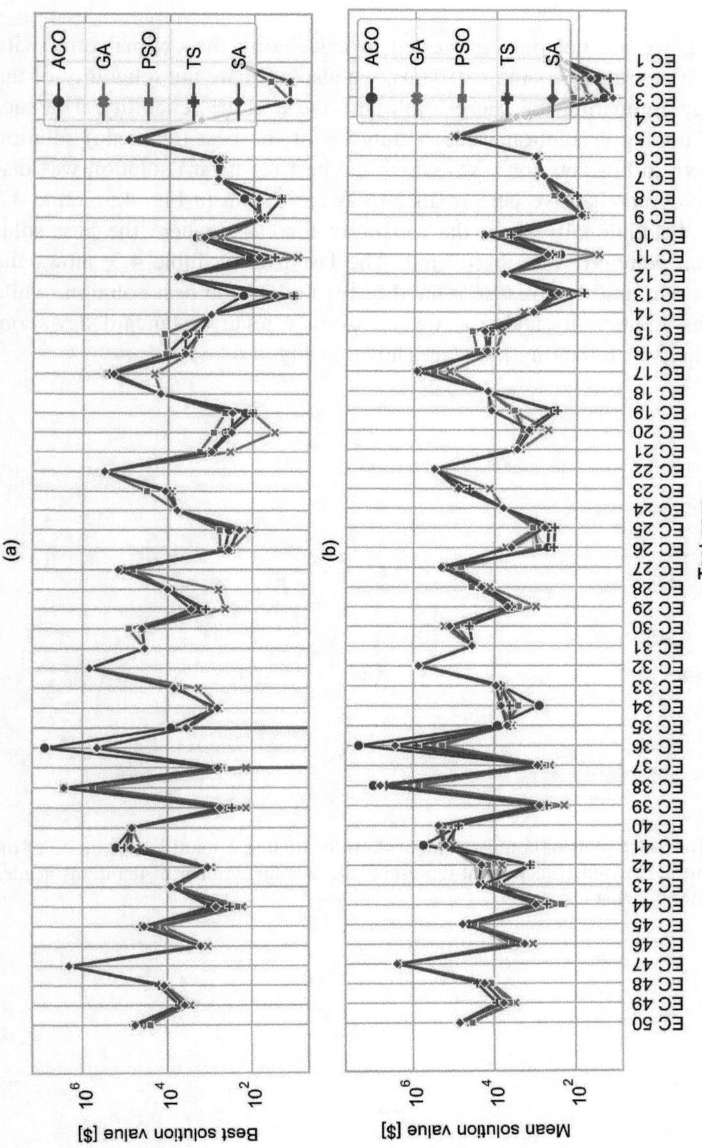

Fig. 4.4 Best (a) and mean (b) solution value for each algorithm for each EC batch, with complexity decreasing from left to right

4.5.2 Reliability: Success Rate and Reliability Plots

As discussed, we are not only interested in calculating the optimal effectivity date once, but constantly monitor it. Thus, we also evaluate the reliability of the algorithms by comparing the success rate and the average reliability. The success rate of finding a solution value within 5% of the best (or mean) solution value is shown in brackets for cases where the best (or mean) solution was discovered. The comprehensive performance plots are shown in Fig. 4.5. Table 4.3 and Table 4.4 additionally show the reliability for cases where the best solution (or mean solution) was discovered. The last row in Table 4.3 shows the total success rate and the average reliability for finding the best solution, while Table 4.4 has additionally added a row for average relative standard deviation. The comprehensive reliability plots are shown in Fig. 4.6.

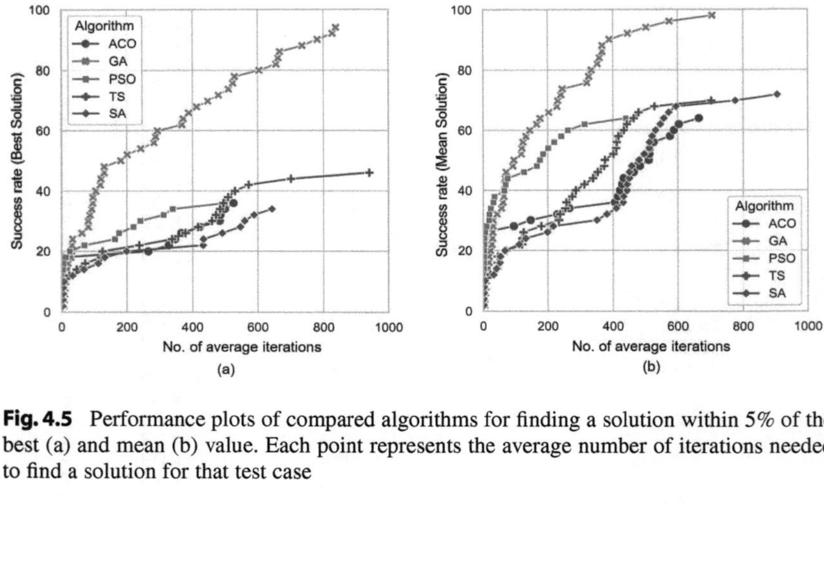

Fig. 4.5 Performance plots of compared algorithms for finding a solution within 5% of the best (a) and mean (b) value. Each point represents the average number of iterations needed to find a solution for that test case

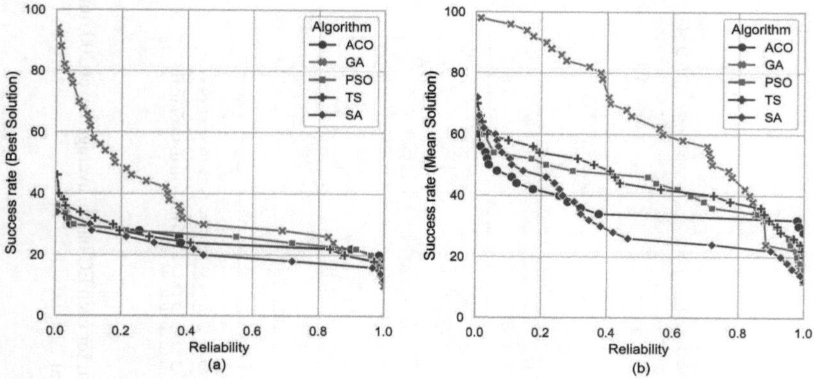

Fig. 4.6 Reliability plots of compared algorithms for finding a solution within 5% of the best (a) and mean (b) value. Each point represents the number of runs where a solution within 5% of the best or mean was found

4.5.3 Speed: Computational Cost

To cover the third criterion the mean time of calculations was evaluated, shown by graphical plots in Fig. 4.7 (a) and Fig. 4.7 (b). Fig. 4.7 (a) presents the absolute time in seconds until the calculations have finished for each EC batch, while Fig. 4.7 (b) shows the values normalised between minimum and maximum. A confidence interval of 95% is shown as the shaded area in both charts. A non-existent shaded area equals low variance.

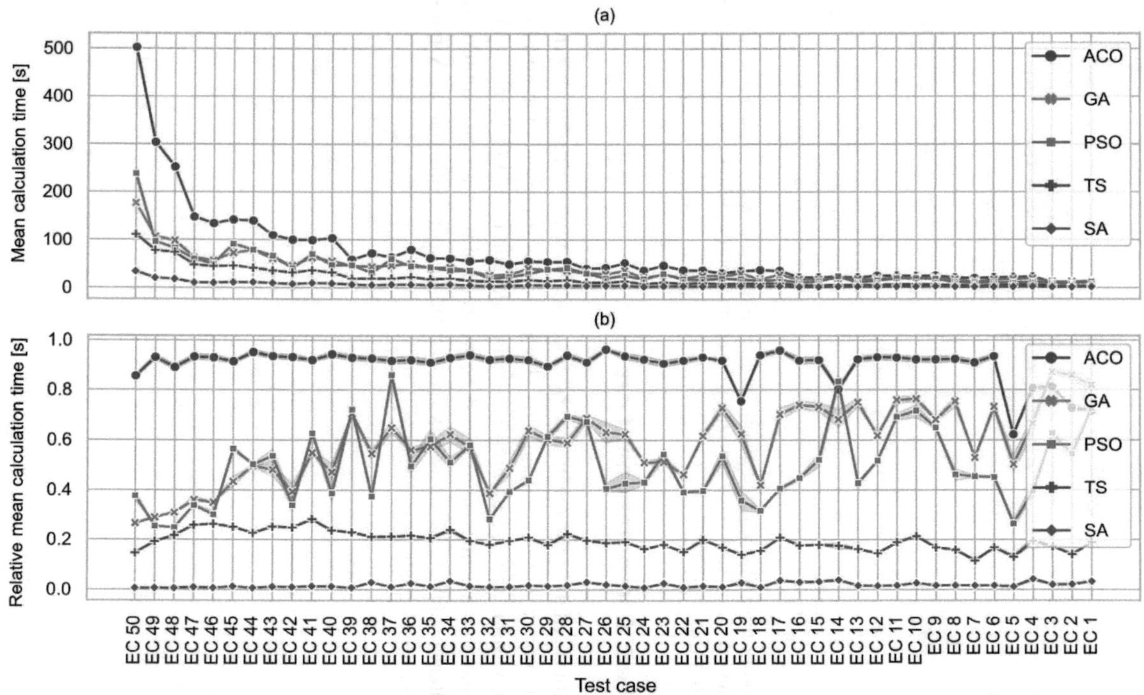

Fig. 4.7 Absolute (a) and relative (b) mean calculation time of a single run for each algorithm for each EC batch, averaged over 200 runs, with complexity decreasing from left to right. Shaded area shows the 95% confidence interval

4.6 Discussion

The model as defined in subsection 4.4.1 proved effective to determine an implementation scenario. The variable costs associated with EC implementation were represented and some general observations can be derived (sect. 4.6.1). A detailed discussion of every algorithm's peculiarity is presented in Sect. 4.6.2, followed by the implications and limitations of our research in sects. 4.6.3 and 4.6.4.

4.6.1 General Observation Regarding Effectivity Dates, Costs, and The Model

4.6.1.1 Analysis of Solution Quality

In general, for 24 out of the 50 cases, the algorithms were able to find a solution which would cost less than 1000$ to execute (Table 4.3 and Fig. 4.3). Out of all cases, only one had a solution which equalled in zero EC implementation costs (EC 36), although upon investigation this should have been possible for ECs 13 and 10 as well as. These changes affected just in-time parts, which were fully consumed on the day of delivery. This results in zero stock at the end of each day. Manually recalculating the cases showed, that an implementation cost of 0 could be achieved by any combination of variables which fit the set of $z_1 = 100$ and $z_2 = z_3$ and any deviation from 100 for z_1 or values for $z_2 > z_3$ violates constraint (5–9). These ideal results were discovered only for one case by the PSO and the GA. A potential reason for this could be the fact that from all possible solutions, only a few would fit this set.

At the other end of solution cost ranges are several cases, which have potentially high execution costs. From all cases, EC 47 stands out (Table 4.3 and Table 4.4) with an expected cost above 2000000$. Upon detailed evaluation, this was due to one of the seven components having a faulty delivery plan in combination with high value goods. One of the components would not see any new delivery until the 33rd production day, although being consumed. As a result, the stock for this component would be permanently negative after the 16th day, violating constraint (4.9). In addition, even with immediately cancelling deliveries of all other components, the remaining inventory on the 16th day would, in combination with high component values, result in high scrapping cost. The same applies to EC 5, EC 38, and EC 43, although at a lower scale.

A similar case is present with EC 41. Here, three components would have a negative stock after the twelfth day, while another low value component would overstock. In combination, this led to the same result: After the twelfth day, no feasible solution exists due to constraint (4.9), while the overstocked number of another component could not be consumed, resulting in increased scrapping costs.

Contrary to the cases before, EC 31, is an EC of low value components. However, more than 60000 pieces are delivered each day. Due to limiting the solution values for z_1 to integers, even if the lowest value of 1 is chosen, 600 components would be delivered, resulting in stock never getting to 0, while having comparatively high penalty costs (from Equation (4.7)). As a result, the algorithms determined immediate stopping of deliveries with an early stopping as a good solution, resulting in increased implementation cost.

EC32, EC30, and EC18 all have consumption costs. As the consumption cost is lower than the sum of each components stock value, it is better to use up all components as much as possible (same as in the example given in Fig. 4.1). As a result, the higher EC implementation costs are a result of the consumption related costs, and not due to scrapping costs.

Finally, EC 27 was a combination of four high-value parts, resulting in high scrapping costs although only a few parts (<5 pieces) are scrapped.

4.6.1.2 Analysis of Effectivity Dates

Within the results, some ECs were scheduled on the same day towards the end of the possible date, while others were comparatively early. Upon further investigation of the production plan, ECs which had an effectivity date on the 88th day were all affecting one specific production facility, which had a production shutdown for the period in question. Thus, any earlier production day would have resulted in high scrapping costs. Additionally, type II EC effectivity dates are impacted by the most expensive part on stock, and its earliest possible new call-off influences the effectivity date of the entire EC batch. Hence, we can conclude that the model proved effective to provide the same results as found in previous research: The optimal date for ECs was as soon as possible close to the earliest possible delivery date of the most expensive new part when running costs (rework, warranty etc.) occur.

4.6.1.3 Analysis of Computation Time

As seen in Fig. 4.7(a), with decreasing complexity, the absolute computation time is reduced as well, although the relative ranking of the algorithm mostly stays the same. Also, it seems that the GA and PSO mostly have a correlating computation time for more complex cases. With reduced complexity, the relative gap to the ACO becomes smaller (Fig. 4.7b). Thus, for simple EC batches the computation time of the algorithm becomes less of a discriminatory factor. The computation time itself is stable for each case. In our runs, the standard deviation of each algorithm was less than 10% of the mean computation time of the specific test case.

Overall, it was shown that our modelling approach is effective for EC effectivity date optimisation. Moreover, the application also revealed errors in the production and demand figures, adding value to monitor production figures. In our model, we have directly consumed tabular production plans and added additional costs. For further model refinement, our model offers increased flexibility compared to previously defined models. More detailed optimisation models could be developed by simply adding more variables and terms to the cost function. From this, future work could research how to include other, more difficult to grasp costs (e.g., complexity costs), and include them in the cost function. However, we do point out that the introduction of every new variable adds computational cost.

4.6.2 Detailed Discussion of the Algorithms

As seen in Fig. 4.5(a), Fig. 4.6(a), and Table 4.3, no algorithm was able to discover a solution within 5% of the best for all test cases, with the GA performing best (47 out of 50 best solutions discovered) and the SA performing worst (14 out of 50). This observation is further supported by the Nemenyi tests in Fig. 4.3, which show that the GA performs significantly better than all other algorithms at finding the best solution. When evaluating the mean cost, the GA's performance is significantly better than the SA's and PSO's. On average, the algorithms converge on the best solution before running the full 1000 iterations (Fig. 4.5). When comparing the success rates and the average reliability across algorithms (last row, Table 4.3, Fig. 4.6), it seems that with increasing success rate for identifying the best solution, the reliability decreases. This behaviour is not observable for mean cost. This is most likely linked to the exploration vs. exploitation balance and indicates that the GA was tuned more towards exploration, while the other algorithms were tuned towards exploitation.

Overall, we can state that the GA performs best at identifying a good solution. As seen in Fig. 4.5(b), compared to the other algorithms, it was most often able to find a solution within 5% of the best solution, with a success rate of 94%. From the evaluated algorithms, it took the second-longest calculation time, with the gap to the ACO closing towards less complex cases (Fig. 4.7b). However, the reliability for reproducing the best result was sometimes lower compared to other algorithms, as seen for instance in EC 7, EC 4, and EC 1. Across all algorithms, it had the worst average reliability (36%) for those cases where a best solution was discovered. Nevertheless, its ability to find the best solution also becomes beneficial for finding a solution better than average. Overall, it found a solution better than the mean in 49 out of 50 cases, with an average reliability of 62%. The only case where it failed to do so, was EC 27. Although its mean solution cost was ranked best on average, with an average relative standard deviation of 1.7% the GA performed second to worst. This leads to the conclusion that the GA, as it was tested, was good for exploring solutions, however its exploitation could be improved.

The PSO was on some occasions able to outperform the GA, requiring less iterations when solutions are found. However, this ability to quickly converge also comes at the cost of a low success rate of 32%. However, compared to the GA, its reliability when an optimal solution is found is comparatively high at 65%, indicating that the PSO's parameters are tuned towards exploitation rather than exploration.

Similar observations can be done for the SA and TS algorithms. They too have comparatively low success rates for the best solution, with a high reliability for those cases where it found one (72% and 58%). When calculating the mean solution cost, the SA algorithm had the highest average relative standard deviations (5.7%) of any algorithm and was ranked lowest on average. Contrary to that, on average the TS algorithm significantly outperforms the PSO and SA algorithms, although it also fails at exploring the search space.

Regarding the performance of the ACO, no significant performance difference can be stated, except that it is significantly worse than the GA at finding the best solution. Furthermore, for more complex cases (number of ECs >3), the ACO requires additional computation time (Fig. 4.7). Providing mediocre solutions consistently matches results in similar research, though the increased computation does not. Upon further investigation, this relates to the specifics of the problem at hand. As the variables are mostly independent of each other, the number of possible paths increases quadratically with each additional date added. Thus, the resulting generated matrix has a size of $(n_{parts} \times n_{variables} \times n_{values})2$, which puts a large strain on memory consumption for more complex cases. The ACO was also unable to improve from the starting condition for the just-in-time ECs (10, 13, and 36). While the other algorithms were sometimes able to find a workaround with $z_2 < z_3$ for these cases, the ACO did not have this possibility due to the heuristic information within the pheromone matrix (cf. sec. 4.4.2.4). Hence, either the heuristic information should be adapted to allow solutions as described above, or for these special cases a different approach needs to be used.

Overall, this benchmark further shows the necessity to test MHs in different environments. Opposed to AHMID ET AL. [146], who have also applied MHs to a combinatorial problem, our research suggests that the ACO is the slowest, and the SA most unreliable algorithm. This is potentially due to the increased search space of the combinatorial problem. While AHMID ET AL. [146] provide a problem with 16 variables and 3 possible values, our problems had up to 123 variables with 101 possible values. This is further supported by the observation, that for simple cases, the ACO performed equally fast, or even faster than the GA, and only slightly slower than the PSO, matching the results in previous publications.

4.6.3 Implications

Two main implications can be drawn from these results:

(1) Of the basic metaheuristics, only the GA is suited to solve most EC effectivity date optimisation problems.
(2) The algorithms struggle with the large search space, necessitating further parameter tuning and research.

The results show that none of the tested algorithms is fully suitable for effectivity date determination, as none has been able to provide the best solution for all test cases consistently. However, the results show that the GA outperforms at exploring possible solutions and provides relatively good solutions on average. It was most often, able to discover the best possible solution. Interestingly, the TS algorithm was able to provide similarly good average solutions, at a fraction of the computational cost. Of the other algorithms, although the ACO's average performance has potential, the computation time took the longest and it failed to compute the solution for several cases. This leads to the conclusion that the ACO is not suitable. Similarly, although the PSO and SA algorithms could identify some best solutions reliably, they could do so only for a fraction of the cases. Furthermore, the variance in the SA's solution cost was underperforming. This translates to the following evaluation of the algorithms, as they were tested:

- The GA algorithm is best at exploring the search space. However, it needs future development to do so reliably.
- The ACO in its current implementation is unsuitable due to an inability to explore solutions. As it also takes too long to calculate, it fails at fulfilling REQ3.
- The PSO algorithm, as it was implemented, was good at exploring the search space, although it converged too fast.

- The TS algorithm was able to provide good solutions on average, although it was rarely the best solution.
- The SA algorithm is unsuitable. It was unable to find good solutions, and its standard deviation of the mean solution was the worst.

The evaluation shows, that no perfect algorithm exists. Additional research is required, for instance, to better tune the parameters. This is made apparent when testing with different values for the σ-parameter. As it indicates the level of exploration, i.e., how much the new solution can vary from the previous one, it is the most important parameter for exploration. In our experiments, we have set this value to ± 15. This means that if the previous value for any of the variables was for instance 50, the next solution values can only be in the interval of [35, 65]. We have run some of the experiments with smaller search spaces of 5 and10, which resulted in higher reliability, but lower success rate for all algorithms and worse solution values. Increasing the search space further would also not be beneficial, as with higher values, heuristic search essentially becomes a random search.

Another implication for implementing the algorithms into production is the computation time. In our experiments, we have limited the number of iterations and compared the wall-clock time. The resulting ratio of computation time from SA:TS:GA:PSO:ACO is approximately 1:3:6:6:10. Translated, when the SA algorithm has run 1000 iterations, the TS algorithm has run 333, the GA and PSO 180, and the ACO has run only 100 iterations. If we now look at Fig. 4.5 and cut off each of the algorithms' performance plots at this number of iterations, the SA algorithm would have solved the same number of test cases as the GA on average (Fig. 4.5b). When it comes to finding the best solution (Fig. 4.5a), however, the GA would still perform best (solving approximately 50% of the test cases after 180 iterations vs. <40% of the SA). However, due to the low reliability of the SA, we would still not suggest using it as a quick optimisation algorithm, but rather use the TS algorithm.

Hence, to operationalise the algorithms, we would suggest a sequential approach, to fulfil all requirements. As the EC effectivity date is determined well in advance (at least 21 days, assumption A7), this initial effectivity date optimisation could be done with running the GA. As it is evident from Fig. 4.5, the GA improves on the solution up until the final iterations. This specialty could be used, to let the GA run until there was no improvement for a given number of iteration cycles (a so-called stopping criterion, e.g., 100 iterations without improvement), or a maximum of wall-clock time has passed. Once the GA has determined an optimal EC implementation plan, the effectivity date should be frozen. This is based on an insight gathered from the industry partner. Once an effectivity date is determined, it is requested by logistics and production planning that it does not change anymore unless a significant economic benefit could be achieved. Any change close to the effectivity date results in increased complexity and production disruptions. Thus, as the search space is now smaller due to the effectivity date being set, the TS algorithm can be monitor and further improve the EC implementation scenario at a fraction of the GA's cost. From there, we expect that the TS can provide similarly good solutions as the search space becomes smaller, and exploration less important. Based on the data shown in Fig. 4.1, the sequential approach could be implemented as:

(1) The GA is initially executed on April 16th (before the timeline shown in Fig. 4.1). This would implicate that deliveries until the 7th of May are off-limits (as it is within the 21-day window).
(2) The GA solution indicates the 23rd of May as the optimal effectivity date, with last deliveries of component 1 on April 25th. Components 2 and 3, due to more frequent deliveries, would be delivered up until May 16th.
(3) As the production figures after the 7th of May are not within the 21-day limit, from April 16th until April 25th the TS algorithm would be run repeatedly to monitor whether the implementation scenario defined on the 16th is still the best and feasible. In case there is a significant change in production figures, and the 23rd of May becomes an unfeasible effectivity date, it would return a warning. Similarly, if it finds a better implementation scenario for components 2 and 3 (e.g., cost saving > 10%) it would return that solution as a better option.
(4) After April 25th, the 21-day window for production figures and the delivery schedule is frozen. After that day, the TS algorithm is run only to control that no supply shortage occurs.

As managerial insight, we have provided a method to consume MRP-II data for optimisation models. Similarly, other data could be used to refine the mathematical model. Furthermore, we have shown that MHs are a quick to adapt tool to identify combinatorial solutions to the EC effective date optimisation problem. However, as an outcome of our research, the basic forms are not fully suitable. Further algorithm and parameter optimisation as well as hybridisation are possible research paths.

4.6.4 Limitations

As limitations, we point out that our research was a blueprint for testing MHs in the EC context, and less a comprehensive assessment of the algorithms. Furthermore, for comparability, we have tested only the base algorithms as they were initially published. Since then, multiple researchers have contributed to improving each one. Future research should hence focus on elaborating and improving the EC implementation model, as well as testing the different variants of the MHs, to see whether some specific variants improve the overall result. Especially increasing the reliability of the GA shows potential for improvement. Furthermore, we have tested the algorithms with a single set of parameters. Modifying the parameters could change the exploration and exploitation performance of the algorithms, but it needs to be tuned toward the deployment environment.

As the test-set is taken from one industry partner, we also suggest testing the algorithms with test-sets from other companies to ensure that they are reproducible. Furthermore, the penalty function for reducing the delivery quantity is heavily dependent on the industry partner and the contractual obligation. Thus, other OEMs could include additional process-related costs for accelerating the transport, tool change, or homologation testing and neglect the delivery costs. As the cost function heavily influences the costs and the computation time, once adaptations are added to the model, the benchmark should be repeated.

4.7 Conclusion

The objective of our study was to compare different algorithms and get an understanding of which algorithm is best suited to solve the EC effectivity date optimisation problem, a variation of the general inventory control problem. For this goal, we have presented a novel optimisation model based on literature and company expertise, which can directly consume production planning data and various costs in EC implementation. Out of all algorithms tested, the Genetic Algorithm was significantly better at discovering the best solutions, combined with a good average performance.

From an industry perspective, however, the low average reliability of the GA indicates that MHs are not yet fully suitable to automate the EC effectivity date problem. Thus, we have suggested a sequential application of the GA for its exploration capability, followed by TS for exploitation and monitoring. To fully operationalise, further development is necessary, and a hybridisation approach as suggested in literature becomes a possible solution path. In future research, these hybrid approaches will be tested, and, if suitable, implemented in a running real-world environment to gather further insight. Eventually, by dynamically controlling the stock quantities during of EC implementation, resource consumption and waste in the industry are reduced.

4.8 Appendix

Table 4.5 Overview EC batches: Number of Components, Factories, and Individual ECs by EC batch

EC batch ID	No. of Components	No. of Factories	type	Individual ECs	Running costs
EC 1	1	1	I	1	
EC 2	1	1	I	1	
EC 3	1	1	I	1	
EC 4	2	1	II	2	
EC 5	2	2	II	2	
EC 6	1	2	I	2	
EC 7	2	1	II	2	
EC 8	2	1	I	2	
EC 9	2	1	I	2	
EC 10	1	2	I	2	
EC 11	2	1	I	2	
EC 12	1	2	I	2	
EC 13	1	2	I	2	
EC 14	2	1	I	2	
EC 15	1	2	I	2	
EC 16	1	2	I	2	
EC 17	1	3	I	3	
EC 18	3	2	I	3	
EC 19	3	2	II	3	
EC 20	3	1	II	3	

(continued)

Table 4.5 (continued)

EC batch ID	No. of Components	No. of Factories	type	Individual ECs	Running costs
EC 21	1	3	I	3	
EC 22	4	1	II	4	X
EC 23	1	4	I	4	
EC 24	4	1	II	4	
EC 25	2	2	I	4	
EC 26	4	1	II	4	
EC 27	4	1	II	4	
EC 28	5	1	II	5	
EC 29	5	1	II	5	
EC 30	3	2	II	5	X
EC 31	2	2	II	5	
EC 32	1	5	I	5	X
EC 33	5	1	II	5	
EC 34	4	1	II	6	
EC 35	2	3	II	6	
EC 36	6	1	I	6	
EC 37	6	1	II	6	
EC 38	1	3	I	6	
EC 39	6	1	II	6	
EC 40	6	4	II	9	
EC 41	4	2	II	9	X
EC 42	7	2	II	9	X
EC 43	10	1	II	10	
EC 44	11	1	I	11	
EC 45	9	2	II	13	
EC 46	7	2	II	13	
EC 47	2	7	II	14	
EC 48	18	3	II	22	
EC 49	16	2	II	28	X
EC 50	41	1	II	41	

Pseudo Code for Algorithms:

Algorithm 1: Ant Colony Optimisation (based on [144])

```
INITIALISATION:
  Initialise Parameters
  Create initial population size N
  Create pheromone matrix
  Evaluate initial population
FOR each iteration DO:
  FOR each ant DO:
    Solution construction using pheromone matrix
    Check if constraints are fulfilled
      IF feasible result THEN:
        Evaluate solution
        Reinforce pheromone matrix
  NEXT
  Evaporate pheromone matrix
NEXT
RETURN: Best solution
```

Algorithm 2: Genetic Algorithm (based on [144])

```
INITIALISATION:
  Initialise Parameters
  Create initial population size N
  Evaluate initial population
FOR each iteration DO:
  FOR k children DO:
    Select two individuals from population
    Create children by crossover
    Mutate children
    Check for constraints
    IF feasible result THEN:
      Evaluate solution
      Add to population
    END IF
    NEXT
  Select new population of N best individuals
NEXT
RETURN: Best solution
```

Algorithm 3: Particle Swarm Optimisation (based on [144])

```
INITIALISATION:
  Initialise Parameters
  Create initial swarm size N
  Evaluate initial population
  Set best solution as swarm best
  Set current solution as particle best
FOR each iteration DO:
  FOR each particle DO:
    Update velocities
    Move to new position
    Check for constraints
    IF feasible result THEN:
      Evaluate solution
      IF solution better than swarm best THEN Update swarm best
      IF solution better than particle best THEN Update particle best
    END IF
  NEXT
NEXT
RETURN: Best solution
```

Algorithm 4: Tabu Search (based on [25])

```
INITIALISATION:
  Initialise Parameters
  Create initial solution
  Evaluate initial solution
  Initialise Tabu list
FOR each iteration DO:
  Initialise new neighbourhood
  Populate neighbourhood with random solutions
  Check for constraints
  IF solution is unfeasible THEN:
    Delete solution
  END IF
  Filter out solutions in Tabu list
  Evaluate solutions
  Set best solution as starting point for next iteration loop
  Update Tabu list
NEXT
RETURN: Best solution
```

Algorithm 5: Simulated annealing (based on [25])

```
INITIALISATION:
  Initialise Parameters
  Create initial solution
  Evaluate initial solution
  Set starting temperature
FOR each iteration DO:
  FOR each temperature holding duration DO:
   Generate random new neighbour
   Check for constraints
   IF solution is feasible THEN:
    Evaluate new neighbour
    IF new neighbour better than previous solution THEN:
     Accept new solution
    ELSE:
     Accept new solution with probability
    END IF
   END IF
  NEXT
  Update temperature
NEXT
RETURN: Best solution
```

4.9 Presentation of Own Contribution

All data presented within the publication was collected and the contents processed by me. This includes the definition of the research scope and research questions, identification of relevant data bases, definition of test sets, identification, definition and application of an optimisation model, and application of the benchmarking method by BEIRANVAND ET AL. [149]. Furthermore, the transformation of the optimisation algorithms into Python code, the execution of the test code and the evaluation of the results was done by me. Additionally, the textual elaboration and writing of the publication were entirely conducted by me. All three co-authors Prof. Dr. Burggräf, Dr. Steinberg, and Dr. Weisser contributed with ideas to the research concept and in the internal review process.

Publication III: Predicting Schedule Adherence of Engineering Changes—A Case Study on Effectivity Date Adherence Prediction Using Machine Learning

5

5.1 Abstract

Engineering changes (EC), redesigns of components, are common with complex products. Their realisation into production systems is a lengthy process and thorough control is imperative. As with most complex processes, however, this is a resource-intensive task and support is required to direct scarce resources. In other domains, predictive business process monitoring (PBPM) research has demonstrated that complex processes can be monitored using machine learning approaches. However, it is indicated that predictive performance is domain sensitive. Therefore, we investigate and show how PBPM can be applied to the EC process by predicting adherence to planned implementation dates. As the outcome of a case study comparing 30 predictive models, our research indicates that EC effectivity date adherence prediction, and thus pre-emptive EC schedule monitoring, is possible. However, performance comparison hints that shallow learning algorithms outperform deep learning algorithms. Furthermore, as the optimal algorithm depends on the deployment scenario, it is demonstrated how cost curves are better decision criteria for choosing models compared to threshold dependent metrics. Based on these findings, this article offers a blueprint

Radisic-Aberger, O., Burggräf, P., Steinberg, F., Becher, A., & Weisser, T. (2025). Predicting schedule adherence of engineering changes—a case study on effectivity date adherence prediction using machine learning. International Journal of Production Research, 63(11), 3913–3937. https://doi.org/10.1080/00207543.2024.2432465

O. Radišić-Aberger, *Optimisation and Control of Engineering Change Schedules in the Automotive Industry with Metaheuristics and Machine Learning*, Findings from Production Management Research,
https://doi.org/10.1007/978-3-658-51074-9_5

for developing machine learning models for predicting the EC schedule adherence and lays the base for further research towards automatic EC scheduling and control.

5.2 Introduction

Engineering Changes (EC) although undesirable are unavoidable, and correctly processing them is a key competitive edge [14]. In the context of manufacturing industries, these ECs can be any modification of the function, behaviour, or structure of a component [4]. ECs span multiple departments of a company and ripple through the supply chain network [16, 54, 110], necessitating orchestration of multiple production locations and products [164]. As a result, their processing is complex by nature. In response, research in engineering change management (ECM) has focused on mitigating negative effects by increasing efficiency and effectiveness in the domain [40], although mostly by statistical methods.

However, with recent advances in artificial intelligence and machine learning (ML), new possibilities emerge to support and automate ECM within organisations [20]. Several publications already discussed ML application in early EC process phases (e.g., [104, 109]), but none of them address the EC implementation phase as a main contributor to EC cost [44]. As delays during implementation cause additional expenses in production for the manufacturer [165], reduced lead-time and adherence to EC implementation schedules becomes a competitive edge. However, absolute control is difficult, as at any time, 1000s of ECs can be running in parallel [11]. To mitigate the administrative burden, which drains capacity from logistics and change coordination departments [26], Radisic-Aberger et al. [1] suggested to research automation possibilities for EC implementation. Here, ML offers opportunities for predicting and monitoring adherence to planned EC implementation schedules, i.e., effectivity dates.

A trending topic in academia and industry, schedule adherence has been successfully predicted by ML researchers in various domains. For instance, Boff Medeiros et al. [166] predicted the length of stay of paediatric patients, Tax et al. [167] predicted whether a loan will be paid back on time, and Burggraf et al. [168] predicted schedule adherence of components ordered for assembly. In all cases, researchers developed ML models to answer the common general question: *Will a case be finished according to schedule?* This was mainly done by applying the technique set known as predictive business process management

(PBPM) [169]. Comprehensive analyses have shown, however, that PBPM performance is process and domain sensitive [170, 171]. Not only does the overall performance of algorithms vary, requiring an evaluation on which is best on a case-by-case base, but moreover, for some business processes the predictive performance is barely better than random guess [171]. Thus, to know whether PBPM methods can support in mitigating the challenge of monitoring effectivity date adherence in ECM, a domain specific investigation is required. Additionally, from a design science research (DSR) [38] perspective, it is necessary to investigate whether the solution is applicable in business.

In this publication we aim to close this gap by investigating how ML and PBPM approaches fare at predicting the adherence to planned effectivity dates. As a research approach, we follow the design science research methodology (DSRM) [172] and develop an information systems artefact. As such, this contribution forms an 'exaptation' (application of known methods in a new domain) and describes the development of a 'Level 2' artefact (showcase of a method) [173]. As no public EC process data set exists [44] the artefact design and development is based on EC data from a German automotive OEM. Likewise, to ensure that the artefact solves the real-world problem, the industry partner was involved in the evaluation.

With the gathered results, our research provides four contributions:

- We show that ML can accurately predict adherence to planned EC effectivity dates.
- We show that shallow learning classifiers outperform deep learning classifiers for EC data.
- We show that cost curves are a superior metric to assess ML model performance under applied business conditions.
- We provide a workflow for model selection in machine learning based prediction of EC effectivity date adherence.

The remainder of this paper is structured according to the DSRM, shown in Fig. 5.1. Following the above stated initial motivation and problem description, Sect. 5.3 infers the objectives of the research by establishing a knowledge base from existing literature. Section 5.4 describes the ML model development, based on the PBPM approach embedded in the common CRISP-DM method for developing and evaluating ML models. The performance results for predicting effectivity date adherence are shown in Sect. 5.5. Section 5.6 deepens the evaluation by discussing these results and a proof-of-concept (PoC) artefact in the business environment. Section 5.7 ultimately provides a summary and outlook.

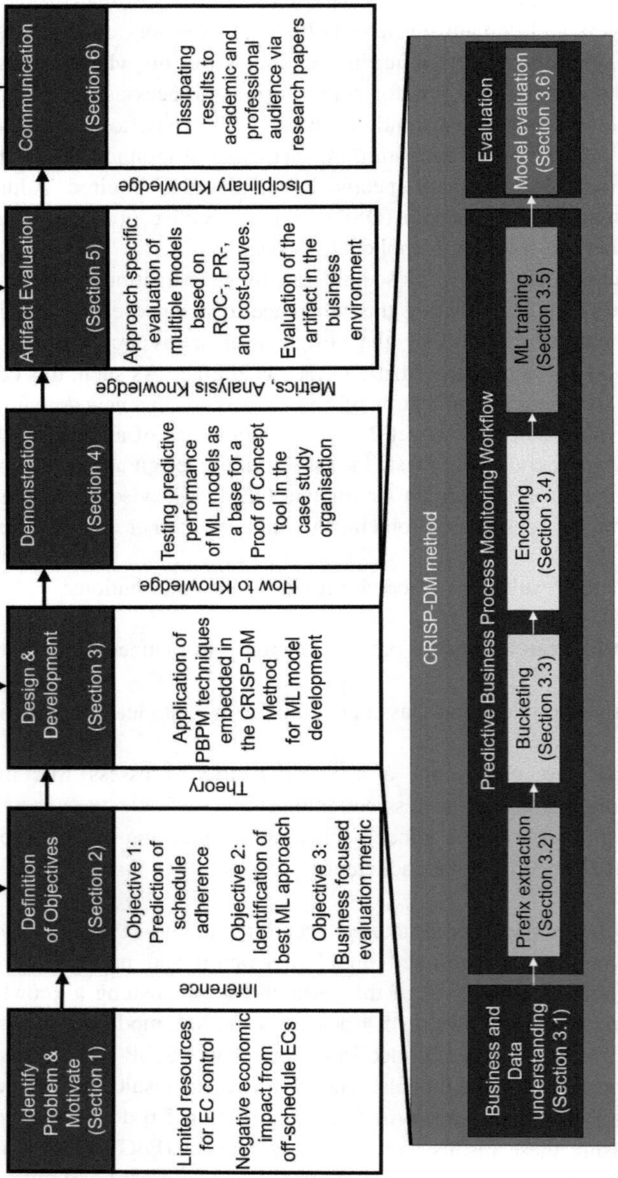

Fig. 5.1 Graphical representation of the research approach. The DSRM Process [172] for the overarching artefact development research, with application of the PBPM [169] workflow embedded in common CRISP-DM framework [103] for ML model development. The respective numbers reference the specific sections in the manuscript

5.3 Background and Related Work

As the second step in the DSRM procedure, literature mandates establishing a knowledge base and therefrom deriving the objectives and research agenda. Within our publication, we therefore further introduce the challenges within EC and the EC process (Sect. 5.3.1), introduce the PBPM methodology (Sect. 5.3.2), and synthesise existing literature and draw a research agenda (Sect. 5.3.3)

5.3.1 Engineering Change and the Engineering Change Process

Although several publications have considered establishing a generic EC process (e.g. [4], [43]) researchers agree that a prescriptive 'one-size-fits-all' approach is not suitable, due to complexity, company, and case specifics [31]. To exemplify, Fig. 5.2 shows a sample of three potential process paths with the general 6-step EC process as described by JARRATT ET AL. [43] shown for reference. The trigger activity is the single commonality across all EC process cases, as it represents its starting point [14] and roughly influences the process path to follow. Ultimately, the EC is realised on its target implementation date, also known as the *effectivity date* [40, 164]. As its meaning varies depending on the perspective of the affected department [55], a research specific definition is required. Since we focus on the effects of late ECs on production systems, we carry over the effectivity date adherence defined by BURGGRÄF ET AL. [164] as *the date on which a pre-change component is used for assembly the last time.*

Ideally, the effectivity date is planned during the approval phase [56] as a function of available stock, running costs and change realisation lead-time [23, 24]. Once the EC has been approved, the EC is released into production and the implementation phase starts [16]. Common activities in this phase are updating purchase orders, ordering post-change components, and performing product and process approval, all managed by a change coordinator [54, 56]. Furthermore, ECs are not introduced independently of each other, but rather in batches [63] or releases [15]. This is done in order to manage the phenomenon of change propagation [13], where changes in one component cause change in other components [14].

However, even with all measures in place, unforeseen problems during the implementation phase such as delayed purchase orders, longer tool lead times, and hidden change propagation [16, 164] can still occur, leading to ECs not being implemented on the planned date [56]. To measure these errors in the implementation phase, CAPISTRANO BURGOS ET AL. [7] introduced the binary key performance indicator (KPI) *batch adherence*. This KPI is a label, displaying 'Adherent' for an EC introduced within its intended EC batch and 'Not Adherent' for all other cases.

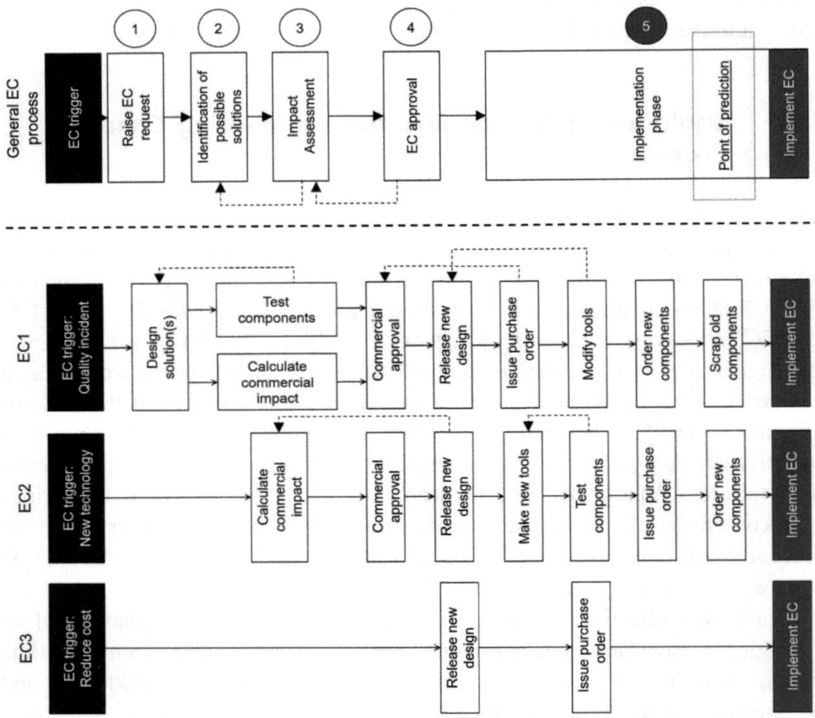

Fig. 5.2 Examples of three different EC process paths depending on the trigger. The loops are exemplary and can occur at any activity. Above, the general EC process as described by JARRATT ET AL. [43] is shown for reference (Steps 1–5, Step 6 'Review of the EC' is left out). Within the implementation phase (Step 5), the approximate process progress at the point of prediction is highlighted to showcase the intended timeframe.

As reports have shown, 'Not-Adherent' ECs can cause multiple complications in assembly systems. In some cases, the cost might be low, such as scrapping obsolete components due to early delivery [23]. However, in severe cases an EC can halt the assembly line if EC batches of post-change parts are delivered in combination with pre-change components [1]. To lower the possibility of late (or early) ECs and improve control, BURGGRÄF ET AL. [44] suggests investigating whether the EC implementation phase can be supported by artificial intelligence. As a starting point, it is suggested to research whether ML can be applied to predict an 'Adherent' or 'Not-Adherent' case, which falls in the research domain of PBPM.

5.3.2 Predictive Business Process Monitoring Methodology

The goal of PBPM is to predict the process outcome for a given case at runtime, be it a numeric variable or a class. Each PBPM project follows a defined workflow, which we base our model development upon. The offline training phase can be described by four major steps [28]:

(1) Prefix extraction
(2) Bucketing
(3) Encoding
(4) Model training

Accordingly, the first task in PBPM is identifying an event log L from which to extract *prefixes* [35]. The event log L refers to a set of documented executions of specific activities, named events e. Event data consists of minimum three components: case ID, activity name, and a timestamp. Table 5.1 gives an example of such an event log for the exemplary case of the three ECs seen in Fig. 5.2. Additionally, case and event payload data can be recorded [35]. While the latter is modified by occurring events, case data generally stays constant throughout a process instance. These two attribute types are hence referred to as dynamic and static data respectively [28]. The sequence of events which refer to the same case ID form a trace σ. To simulate this the prediction of ongoing process during ML training, the completed traces $\sigma = [e_1, \ldots, e_n]$ of length n are truncated to a $prefix(\sigma, l) = [e_1, \ldots, e_l]$ so that $l \leq n$ [35]. The term *prefix* thus subsumes all data which constitutes the trace until the point of prediction.

Table 5.1 Example of an event log based on the initial steps of the EC process cases as seen in Fig. 5.1

Case ID	Activity	Timestamp	Payload
EC 1	EC create	1.9.2022	quality
EC 2	EC create	2.9.2022	technology
EC 1	Design solution	2.9.2022	–
EC 3	EC create	3.9.2022	cost reduction
EC 1	Commercial approval	10.9.2022	approved

In a second step, to amplify predictive performance of ML models, the extracted prefixes are split into groups of similar cases, called buckets. This bucketing step is optional, although often shown to be effective in improving predictive performance as bucket-specific models are not expecting features exclusive to certain cases [28]. This is either done automatically by applying a clustering algorithm (e.g. K-Nearest Neighbour), by trace length, or manually based on domain knowledge.

After bucketing, each prefix execution trace σ_i is encoded as a feature vector $g_i = (g_{i1}, g_{i2} \ldots g_{ih}, label_i)$ [28]. In comprehensive studies it has been shown that neither the simplest nor the most sophisticated encodings deliver the best result, and a middle ground should be used, e.g. aggregation or simple index-based encoding [35, 170]. An aggregation-based example derived from Fig. 5.2 is presented in Table 5.2. In addition to the explicit recorded information of an

Table 5.2 Example for aggregation-based encoding, based on the three ECs shown in Fig. 5.1. Each occurrence of an activity is counted and aggregated

Case ID	EC trigger	Design sol.	Test comp.	Calc. comm. impact	Comm. appr.	Rel. new design	Issue purch. order	Make new tools	Modify tools	Order new comp.	Scrap old comp.	Eff. Date Adh.
EC 1	1	2	2	1	2	3	2	0	2	1	1	1
EC 2	1	0	2	2	2	2	1	2	0	1	0	0
EC 3	1	0	0	0	0	1	1	0	0	0	0	1

event, literature suggests enriching the models with synthetic data, e.g. weekday information, time distance features, or complexity representation [28].

The resulting feature vectors are subsequently applied as training data to ML algorithms [28]. Common supervised algorithms in PBPM are Random Forest (RF) [174], XGBoost (XGB) [175], and long-short-term-memory (LSTM) neural networks (NN) [176], as seen in [167, 170, 177]. Both, RF and XGB are referred to as shallow learners built on ensemble learning, however, their inner structure differs. RF consist of a random set of independent decision trees and each unique output is merged by majority voting. Compared to that, at the core of XGB lies an ensemble of weaker prediction models, which are developed in sequence and learn from the previous level. NN, such as the LSTM network on the other hand, mimic the human brain with an input layer, an output layer, and an n-number of hidden layers in-between. Within its architecture, each input cell is sharing its information with each cell of the next layer and so forth, until an output is generated. For a detailed description, we refer to the initial publications introducing these algorithms.

Furthermore [171] showed that the performance of deep learning (DL) models outperforms shallow ML-models in case of high imbalance of target variables, high activity-to-instance ratio (i.e. cases in which most data is generated at runtime), and high variant-to-instance ratio (i.e. cases in which only a small number of processes follow a standardised process path). All three attributes apply to the EC process, suggesting that DL algorithms perform better at predicting the EC process outcome.

Once the models are trained, the best performing model is operationalised for predicting an unfinished new case. However, PBPM research is inconsistent regarding the best metric for model evaluation. Some researchers apply accuracy, F1-scores, or the area under the receiver-operator-characteristics curve (AUC-ROC), or all of them as a decision metric [35, 171].

To summarise, predicting the outcome of business processes such as the EC process falls in the application domain of PBPM. In a usual research environment, model development consists of four phases—prefix extraction, bucketing, encoding, and model training. Of the common algorithms tested, XGB, RF and LSTM networks have excelled in past research. However, the EC domain is specifically oriented towards business needs, which is why the common metrics are not enough.

5.3.3 Predicting Adherence to Engineering Change Schedules

Based on the discussions of EC and PBPM theory, we can deduce three objectives which should be fulfilled by the resulting artefact (2nd Step of the DSRM, Fig. 5.1):

- Objective 1: Prediction of schedule adherence
- Objective 2: Identification of best ML approach
- Objective 3: Business focused evaluation metric

However, in the EC domain currently no known publication exists addressing explicitly the EC schedule prediction problem, nor all objectives. Nevertheless, it does offer approaches for EC process and EC implementation optimisation, which could be applied for schedule adherence improvement. Firstly, simulation tools, such as the resource constrained optimisation approach by LI ET AL. [121] and the Monte-Carlo simulation-based approach by WYNN ET AL. [178], estimate the duration of design activities and calculate an expected implementation time. However, simulations are too expensive and not feasible on a large scale. Similarly, the qualitative approaches by WÄNSTRÖM ET AL. [26] and STEKOLSCHIK [27] are not satisfactory. Here the change coordinator should focus on specific ECs based on expected economic impact. While an alleviating strategy at first, ultimately this results in inefficiencies due to doublechecking without necessity, while accepting that some ECs are off schedule due to low impact. Additionally, most AI-driven ECM research has only investigated predictive performance from an academic perspective, without reporting on the implementation in industry [44].

From the wider area of PBPM research and schedule prediction, several approaches exist for a specific domain or on a general level. Coming from a production perspective, BURGGRAF ET AL. [168] discuss how ML can be applied for predicting late parts from suppliers. In their research, they applied various algorithms to a classification task (late order delivery), showing that AdaBoost and Gradient Boost algorithms outperform other shallow learners (RF and linear regressor) and a DL multi-layered perceptron NN. However, the process under investigation (order placement) had no loops and was strictly sequential, limiting the transferability to complex processes such as EC. Research within the medical domain for schedule adherence was conducted by BOFF MEDEIROS ET AL. [166] who showed that RF outperforms a support vector regressor. However, their process was rather reduced, as the only process flow related data points were time of admission.

Table 5.3 Overview of literature on existing methods for EC schedule adherence optimisation

Contribution		Applied method	**Objective 1: Schedule adherence prediction**	**Objective 2: Identification of best ML approach**	**Objective 3: Business focused evaluation**
ECM Domain	[26]	Decision tree prioritisation method based on effort and expected loss	-	-	✓
	[110]	ML based identification of lead time influences	-	✓	-
	[178]	Monte Carlo based workflow simulation.	~*	-	✓
	[27]	Effort based decision model for setting organisational focus	-	-	✓
	[121]	Resource Constrained scheduling optimisation	~*	-	-
	[109]	ML for change propagation during EC process	-	✓	✓
	[104]	ML for EC effort prediction	-	-	-
Domain specific	[168]	ML for late order prediction	✓	✓	-
	[166]	ML for predicting length of stay	✓	✓	-
PBPM research (Domain independent)	[167]	LSTM for predicting remaining time in process	✓	✓	-
	[179]	Comparison of ML application and cost models	-	-	✓
	[35]	Comparative PBPM study of LSTM, XGB and RF	✓	✓	-
	[171]	Comparative PBPM study of LSTM, SVM, DNN and RF	-	✓	-

ML = Machine Learning, LSTM = long-short-term-memory neural net, SVM = Support Vector Machine, XGB = XGBoost, RF = Random Forest, DNN = Distributed Neural Net
* While a simulation can predict the adherence of EC effectivity dates, it does so by calculating potential paths and durations. This is possible for single cases, but not suitable for monitoring large amounts of ECs.

General PBPM literature compares various ML approaches for their performance, however, seldomly from an industry perspective. For instance TAX ET AL. [167], TEINEMAA ET AL. [170], TEINEMAA ET AL. [179], and KRATSCH ET AL. [171] provide comparative reviews of RF, XGB, LSTM, and other ML algorithms for outcome prediction. By focusing on identifying new techniques and testing across multiple available benchmark data sets, their research neither discusses business application, case specifics, nor domain specifics. Only KRATSCH ET AL. [171] further discusses how data set specifics could influence the performance of DL and shallow learning algorithms. Finally, TEINEMAA ET AL. [179] discusses the necessity of a business related decision metric for choosing the ideal algorithm. Therefore, they design a metric based on costs of false predictions. Their research shows how the cost-optimal model is threshold dependent. However, they only compared a RF and a Gradient Boost model. Furthermore, it is unknown how the approaches presented fare in the EC domain.

In conclusion, none of the existing approaches, addresses all objectives (Table 5.3) stated. Thus, it is necessary to first test whether ML algorithms can predict the adherence to effectivity dates, and secondly decide which algorithm does so best. To do so, we generalise the concept of batch adherence to '*effectivity date adherence*': It is 'Adherent' for all cases where the effectivity date is within the planned timeframe, and 'Not Adherent' for all off-schedule ECs. Third, we expand on the common performance metrics with cost curves [180], a business sensitive evaluation metric.

For study guidance, the objectives can be translated into three research questions (RQs) which we intend to answer with our research:

- RQ1: How well can supervised ML predict the adherence to EC effectivity dates?
- RQ2: What metric is applicable to determine the best model to apply?
- RQ3: How to determine economic usefulness of applying supervised ML?

In summary, we focus on comparing ML-models performance regarding predicting the adherence to planned effectivity dates, labelled 'Adherent' or 'Not Adherent'. To achieve this, we intend to use static data, e.g. descriptive EC data, as well as dynamic data which emerges from the EC process and the activities within. As a main decision criterion, we compare the predictive performance of ML approaches in a business context.

5.4 Materials and Methods

As the third step in the DSRM framework, the design and development shall be presented. For this step, we employ the PBPM method [28] as described in Sect. 5.3.2. However, for applicability of the PBPM methods, they must be enriched with common ML steps: Firstly, business and data understanding, secondly the evaluation of the developed methods. Therefore, we embed the PBPM method in the CRISP-DM framework [103], replacing the mandated data preparation and modelling steps (Fig. 5.1). For easier reference, we have named the respective sub-sections according to these steps.

5.4.1 Business and Data Understanding

5.4.1.1 Business Understanding

Within ECM research, no public data set exists for testing and training algorithms [44]. Thus, a case-based approach was chosen, with real world data from an automotive OEM. The collected data consists of ECs which were started and finished in the timespan from January 1st, 2021, until June 31st, 2022. Older data was not considered as the EC process saw major modifications in the approval system before, and newer cases were not yet finished. Within these 18 months, the company has seen major product launches and has dealt with regular day-to-day changes. Thus, it can be noted that a wide variety of changes are part of the data set.

The company has implemented a holistic EC batching strategy to reduce change complexity. As part of the modification, the ECM system was upgraded towards a comprehensive product data management system. Since then, it integrates all relevant departments and ensures that all data and EC related activities are recorded in detail and automatically. Already before the ML study, process mining activities have been implemented for thorough optimisation. To ensure that adverse timing effects during EC implementation are reduced, supply chain change coordinators must confirm the planned effectivity date with the respective suppliers before EC approval. Likewise, tasks such as purchase order filing and product and process approval is all recorded and shared within the ECM system.

Furthermore, the OEM has implemented an effectivity date adherence KPI, which is built upon logistical data recorded by the manufacturing resource planning system. It translates to the definition given in Sect. 5.3, where it records the incoming and consumed components at the assembly line, including their change version. In case it is consumed after the planned implementation date, the specific EC is automatically flagged as 'Not Adherent'. If no pre-change component is used after the implementation date, and a post-change component is consumed at least once, the EC is flagged 'Adherent'. All non-flagged changes are considered 'to be implemented / open' and not part of our data set.

As part of the case-based research, several discussions with staff from the OEM were conducted. This included senior management from ECM and design departments and their associates, as well as purchasing, supplier quality assurance, and logistics departments. These talks were not according to a predefined questionnaire, but rather open talks to gather data and business understanding (Steps 1 and 2 of the CRISP-DM method. Whenever we mention 'expert talks' in the rest of the manuscript, we refer to these groups of individuals mentioned above. As a result of these talks, we also determined the ideal time of prediction: three weeks before the first delivery date deadline. The reasoning is linked to the supply chain, as within the three-week period a call-off cannot be changed without commercial penalty. Thus, any change to the delivery schedule before can be considered cost-free, which translated to a good prediction point.

5.4.1.2 Data Understanding

For gathering the event log, data mining focused solely on the available ECM execution system. This excludes more detailed data from supporting systems such as procurement or quality assurance systems. The full data set consists of 62 379 unique completed ECs, which were introduced in 5685 batches. An EC was considered completed if it was flagged as 'Adherent' or 'Not Adherent' as defined in sub-section 5.4.1.1 The entire event log consists of 1 115 721 entries.

Further high-level investigation of the data log provides information necessary for comparing the research outcome to other publications. Due to the domain comprehensiveness of PBPM, we describe the data set according to KRATSCH ET AL. [171]. Starting with the sequence analysis, the 62 379 ECs were processed in 6505 unique sequences. The most common sequence was used with 517 ECs, 970 ECs had a unique sequence of events. However, across batches, we can see that there are more EC process paths than batches. The most common process path was taken by 97 EC batches, whereas 5972 EC batches had unique process paths. Thus, the instance-to-variance ratio has two answers: The primary variant-to-instance ratio is 9.59, indicating a rather repetitive process. However, if the EC batch is considered as the instance level, then there are more process paths than instances, resulting in a value of 0.87 showcasing the complexity of EC. In total, 60 different activities were recorded before data cleaning. The longest process instance saw 94 events, the shortest processes were finished after 2 recorded events (excluding the implementation activity), averaging at 18. On average, each EC instance saw 14 different activities, with a maximum of 28. With a resulting event-to-activity ratio of 14/18, this indicates repetition and frequent loops in the process. The overall positive class ratio was 0.24 (minority class, 'Not adherent'). Regarding the data types and cardinalities, these are displayed for the final features in Table 5.5 and Table 5.6.

5.4.2 Prefix Extraction

According to the PBPM workflow, prefix extraction is the first step in any project. In our research, this step compliments the data understanding step and initiates the data modelling step from the CRISP-DM method. Within this step, PBPM literature suggest cleaning the data, conducting feature engineering tasks, and selecting a base feature set used for ML.

For the initial set of features, all available data from the ECM system was considered. In a first step, non-generalisable features were removed, as they hold no value for future predictions. This was sometimes rather apparent, e.g., specific dates (they will never occur in the future), but also affected some high cardinality features, e.g., part numbers, model codes and supplier data. While part numbers have been omitted without further discussion (cardinality >10 000), model codes and supplier data were discussed thoroughly with the associates at the OEM. In the end, model codes were kept as a feature as they are available for multiple years and have some generalisable aspects. Manufacturer specific data was also discussed thoroughly. However, in the end we only considered the manufacturer country as a relevant feature. The supplier code, i.e., the manufacturer name, was dismissed, as it would have a high cardinality (>1000), with little information for the future. Mainly from expert talks, we gained the understanding that a good or bad performing supplier in the past might change his performance in the future.

Afterwards, thorough data cleaning was undertaken. Features, which were only occurring with less than 5% of the ECs were dismissed. Missing information on categorical data was interpreted as its own category ('*missing*'). Categorical values (except for activity description) occurring with less than 5% of ECs were grouped as '*other*'. Features, which only existed for older or newer ECs (e.g., because they became obsolete when the EC process was modified), were removed or replaced with a successor if available. Following a correlation investigation, pairs of features which had a high correlation (>0.7 on a spearman rank correlation) had one feature removed until every remaining correlation pair was below 0.7. Ultimately, based on the talks with experts, 22 static and ten dynamic features were kept.

For the specific activities, the same processing was undertaken but on feature level, e.g., activities which occurred with less than 5% of ECs were dismissed. Activities, which occurred only for a specific time frame (e.g., first three months of the observation time frame) were either dismissed, or if available linked with a successor activity. Missing information on activity end was considered as a non-occurrence. Thus, from the initial 60 activities a total of 31 was kept (shown as dynamic categorical feature 'Activity Description'). All 31 activities had been automatically recorded by and were afterwards extracted from the ECM system. In accordance with the prediction strategy (ref. Sect. 5.4.1), only activities up to three weeks before first planned delivery were considered in the data set. Ultimately, the traces were truncated to a prefix length of 30, as 90% of the traces of the full dataset have completed by then.

5.4.3 Bucketing

Instead of using an automatic unsupervised bucketing approach, we used domain knowledge bucketing based on the EC trigger. The trigger is indicated by two features, *quality relevant EC* and *commercially relevant EC*, and determines the minimal process path. If an EC is neither, a purely design driven EC process is indicated. Four trace buckets resulted from this. To compare the effectiveness of the bucketing approach, the algorithms were tested also on the unified data set consisting of all recorded cases. Table 5.4 summarises the resulting dataset statistics.

Table 5.4 Dataset statistics

Dataset	Bucketing Logic	# traces (ECs)	# batches	pos. class ratio[a]	Average prefix length	# Static features	# Dynamic features[b]	Dimension[c]
Dataset Full	All	62 379	5685	0.24 (0.22)	18	32	11 (22)	123 (1174)
Dataset 0.0	¬C ∧ ¬Q	36 878	3675	0.26 (0.23)	14	25	10 (21)	112 (1105)
Dataset 1.0	C ∧ ¬Q	18 755	1077	0.24 (0.16)	16	26	10 (20)	110 (1071)
Dataset 0.1	¬C ∧ Q	3651	151	0.18 (0.45)	17	21	9 (19)	102 (885)
Dataset 1.1	C ∧ Q	3095	201	0.17 (0.19)	17	30	11 (22)	105 (905)

Q = Quality, C = Commercial, ¬ = logical 'Not' Operator, ∧ = logical 'And' Operator,
[a] Percentage of positive process outcome class ('Not Adherent') for training data (test data ratio shown in brackets)
[b] Event specific features (e.g. time since last event) are only available for index-based encoding (shown in brackets)
[c] Dimensionality after aggregation encoding and index-based encoding (shown in brackets)

5.4.4 Encoding

The third step in PBPM is the encoding of the feature vectors. As the different encoding techniques enable different synthetic features, the encoding phase is extended with feature engineering.

In total, 22 synthetic static and dynamic features were created, based on the data, literature, and expert talks. These synthetic features were mainly time spans deduced from the timestamps of activities, calculated as described in [181]. Furthermore, based on the factors affecting EC implementation presented in [6], we synthesised the overlap of project stages by counting the number of affected models and factories, as well as the experience of the supplier by calculating how many ECs were executed in the past. Three features were synthesised to display the complexity of an EC batch: Number of ECs within the batch, number of affected plants, and number of affected models. Table 5.5 and Table 5.6 show the full list of static and dynamic features with additional meta information, such as data type and cardinality of categorical features.

Table 5.5 Static features used for predicting the outcome of the EC process

Feature	Possible Values	Level[f]	Description
Static Numerical Values			
# previous quality ECs	{0, ..., n}	EC	indicates how many EC were already implemented to fix a quality issue
# concurrent ECs per batch	{0, ..., n}	Batch	Indicates how many ECs are grouped into the same EC batch
Project maturity	{0, ..., n}	EC	numeric representation of how far the project has progressed in a products lifecycle, with max value = series production
Quality severance	{1, ..., 9}	EC	numeric representation of the severity of a quality issue, with 1 being life threatening issues, 9 being no issues recorded
Quality importance	{0, 1, 2}	EC	numeric representation whether the quality issue needs special attention (2), or whether it can be solved by the production crew themselves (0)
# affected models[e]	{1, ..., n}	Batch	indicates how many products are affected by the EC batch
# affected factories[e]	{1, ..., n}	Batch	indicates how many factories are affected by the EC batch
Manufacturer Experience[e]	{0, ..., n}	EC	indicates how many EC batches were already executed by the supplier
Creation date → commercial EC approval deadline[e]	$\{-\infty, \ldots, +\infty\}^c$	Batch	indicates how many days before the commercial approval deadline the EC batch was initiated
Creation date → last possible Effectivity Date[e]	$\{-\infty, \ldots, +\infty\}^c$	Batch	indicates how many days before the last preferred effectivity date the EC batch was initiated
Creation date → planned purchase order date[e]	$\{-\infty, \ldots, +\infty\}^c$	Batch	indicates how many days before purchase order availability deadline the EC batch was initiated

(continued)

Table 5.5 (continued)

Feature	Possible Values	Level[f]	Description
Creation date → planned Effectivity Date[e]	$\{0, \ldots, +\infty\}^c$	Batch	indicates how many days before the planned effectivity date deadline the EC batch indicates initiated
Creation date → eng. EC approval deadline[e]	$\{-\infty, \ldots, +\infty\}^c$	Batch	indicates how many days before the engineering approval deadline the EC batch was initiated
Creation date → earliest possible Effectivity date[e]	$\{-\infty, \ldots, +\infty\}^c$	Batch	indicates how many days before the earliest preferred effectivity date the EC batch was initiated
Creation date → first delivery date deadline[e]	$\{-\infty, \ldots, +\infty\}^c$	Batch	indicates how many days before the new parts delivery deadline the EC batch was initiated
Static Categorical Values[a]			
Cost increase EC	{1,0}	Batch	indicates whether the EC batch increases the projected budget costs
Cost decrease EC	{1,0}	Batch	indicates whether the EC batch decreases the component production costs
Initial release EC	{1,0}	Batch	indicates whether the EC batch is used for an initial release of new components
Warranty EC	{1,0}	Batch	indicates whether the EC batch fixes a warranty case
Software relevant EC	{1,0}	Batch	indicates whether the EC batch affects mechatronic components
Documentation EC	{1,0}	Batch	indicates whether the EC batch fixes missing documentation
Commercially relevant EC	{1,0}	Batch	indicates whether the EC batch is commercially relevant
Quality relevant EC	{1,0}	Batch	indicates whether the EC batch is relevant for an unspecified quality issue

(continued)

Table 5.5 (continued)

Feature	Possible Values	Level[f]	Description
Delivery to external warehouse	{1,0}	EC	indicates whether the component is supplied via an external logistical centre to the factory
Follow up EC	{1,0}	Batch	indicates whether the EC batch fixes a previous EC
EC scenario (Cardinality 2)	{1,2}	EC	indicates whether the EC is a scenario I or II case
Planned purchase order set manually	{1,0}	EC	indicates whether the planned purchase order date deviates from the generically predetermined
EC during new product development	{1,0}	EC	indicates whether a new product is affected or a running series product
EC process type (Cardinality 3)	{AP, CP, GP}	Batch	indicates which process components are active for the EC batch
EC batch type (Cardinality 3)	{A, BiW, HV}	Batch	indicates which activities to execute based on assembly, body in white, and high voltage components
Factory (Cardinality 12)	{1, 2, ...,12}	EC	indicates which factory is affected by the EC batch
Manufacturer country (Cardinality 18)	{DE, CZ, ...}	EC	indicates the production location of the supplier of each component, values occurring less than 5% are grouped in "other"
Detailed EC type (Cardinality 11)	{FC, GC, ...}	EC	indicates the detailed EC type, e.g., drawing change, functional change etc.

[a] Categorical feature values that occurred in less than 5 % of cases were grouped as 'other'
[b] These features are triggered or modified by an activity, thus called payload
[c] $-\infty$ can be possible, if the EC batch is initiated after the deadline for a specific project
[d] $-\infty$ can be possible in cases where specific activities are executed before the initiation of the EC batch
[e] synthetic feature
[f] some events are EC specific rather than applicable for the entire EC batch

Table 5.6 Dynamic features used for predicting the outcome of the EC process

Feature (Value Type)	Possible Values	Level[f]	Description
Dynamic Numerical Values			
Event date → last possible Effectivity Date[e]	$\{-\infty, \ldots, +\infty\}^{c}$	Batch / EC	indicates how many days before the last preferred effectivity date the activity has occurred
Event date → creation date[e]	$\{-\infty, \ldots, +\infty\}^{d}$	Batch / EC	indicates how many days have elapsed since the initiation of the EC batch
Event date → engineering EC approval deadline[e]	$\{-\infty, \ldots, +\infty\}^{c}$	Batch / EC	indicates how many days before the EC approval deadline the activity has occurred
Event date → first delivery date[e]	$\{-\infty, \ldots, +\infty\}^{c}$	EC	indicates how many days before new parts delivery deadline the activity has occurred
Event date → planned purchase order date[e]	$\{-\infty, \ldots, +\infty\}^{c}$	EC	indicates how many days before the purchase order availability deadline the activity has occurred
Event date → earliest possible Effectivity Date[e]	$\{-\infty, \ldots, +\infty\}^{c}$	EC	indicates how many days before the earliest preferred effectivity date the activity was executed
Event date → planned Effectivity Date[e]	$\{0, \ldots, +\infty\}^{c}$	EC	indicates how many days are left until the target effectivity date when the activity was executed
Activity completion delay[e]	$\{-\infty, \ldots, +\infty\}$	Batch / EC	indicates whether the activity was delayed
Activity duration[e]	$\{0, \ldots, +\infty\}$	Batch / EC	indicates how long it took to execute the activity
Time elapsed since last event[e]	$\{0, \ldots, +\infty\}$	Batch / EC	indicates how much time has passed since the last activity
EC process progress[b]	$\{1, \ldots, n\}$	Batch / EC	numeric representation how much the EC process has progressed, with max value = finished

(continued)

Table 5.6 (continued)

Feature (Value Type)	Possible Values	Level[f]	Description
Commercial approval progress[b]	{1, …, n}	Batch	numeric representation how far the commercial approval process has progressed, with max value = approved
Predicted logistical costs[b]	{0, …, + ∞}	EC	indicates the predicted costs for scrapping old parts
# assembly trials conducted[b]	{0, …, + ∞}	EC	indicates how often the new component was tested in an assembly trial run
# of events[a]	{0, …, + ∞}	EC	indicates how many events have occurred for the EC
Dynamic Categorical Values[a]			
Activity has target date (Boolean)[e]	{1,0}	Batch / EC	indicates if the activity itself has a deadline
Activity description (Cardinality 31)[e]	{OP, RD, …}	Batch / EC	description of the activity, e.g., order parts, release drawing, etc.
Purchase order available (Boolean)[a]	{1,0}	EC	indicates whether a purchase order is available, payload triggered by the corresponding activity
logistically relevant EC (Boolean)[a]	{1,0}	EC	indicates whether an EC is affecting logistics, payload triggered by the corresponding activity
Confirmed delivery (Boolean)[a]	{1,0}	EC	indicates whether the material planner has confirmed delivery, payload triggered by the corresponding activity
Confirmed Supplier Readiness (Boolean)[a]	{1,0}	EC	indicates whether the supplier has confirmed delivery, payload triggered by the corresponding activity

(continued)

Table 5.6 (continued)

Feature (Value Type)	Possible Values	Level[f]	Description
Confirmed EC coordination status (Cardinality 6)[a]	{E, E & P, …}	Batch	indicates the level of confidence of the change coordination board, e.g., engineering confirmed, engineering & production confirmed

[a] Categorical feature values that occurred in less than 5 % of cases were grouped as 'other'
[b] These features are triggered or modified by an activity, thus called payload
[c] $-\infty$ can be possible, if the EC batch is initiated after the deadline for a specific project
[d] $-\infty$ can be possible in cases where specific activities are executed before the initiation of the EC batch
[e] synthetic feature
[f] some events are EC specific rather than applicable for the entire EC batch

As prefix encoding techniques, we decided to perform experiments on aggregation and simple index-based last-state encoding [28], as these have led to the best results in past research [170, 177, 182]. The RF and XGB algorithms were provided with an aggregation encoded dataset and a flattened index-based data set (i.e., each event and payload data of a trace was added as an additional column). The DL algorithms were only given the index-based last-state data according to [170]. Lastly, categorical features were one-hot-encoded.

In total, the resulting aggregation encoded data set had a dimensionality of 129 (123 encoded features, 5 IDs and keys, and 1 target label) and the simple index-based data set a dimensionality of 1180 (1174 encoded features, 5 IDs and keys, and 1 target label), mainly inflated by adding the activity description of the 30 recorded events. Afterwards, standard ML pre-processing such as min-max scaling (for finite integer ranges from 0 to n) and mean value standardisation (for infinite ranges) was applied. Oversampling or undersampling, sometimes useful to deal with imbalanced datasets, was left out as it showed no performance increase during the exploration phase. Similarly, principal component analysis was tested during the exploration phase, however, it showed no improvement and was omitted subsequently.Please note that the Table 5.5 indicator 'b' and 'd' footnote text is mentioned below. However, citations are not included in the table. Please check and make any necessary changes.Thank you for pointing this out. As Table 5.5 and Table 5.6 show similar information, it was decided to have for both tables the same indicators for easier readability.

5.4.5 Model Training

As a starting point, we chose to test the RF, XGB, and LSTM algorithms, as they have shown good results in other PBPM research. Additionally, coming from a text-based use case LICHTENWALTER ET AL. [183] showed that branching neural nets (BNN) with specialised layers can outperform unimodal LSTM NN. Thus, we compared four algorithms: RF, XGB, as state-of-the-art shallow ML-models for PBPM, an LSTM NN as described in [170], and an adapted version of the BNN by LICHTENWALTER ET AL. [183].

RF and XGB are readily available algorithms from the Scikit Learn and XGBoost library for Python, whereas the NNs were implemented using the Keras library. For the detailed code implementation of the LSTM network, we refer to [170] (adaptation for classification) and [167] (base network). The BNN was adapted for our use-case. As LICHTENWALTER ET AL. [183] were using a text-based semi-structured data set, they used embedding layers for categorical features. However, as none of the features in the EC data set had a high cardinality, instead of an embedding layer we used a multi-layer perceptron for categorical and numerical features (i.e., static data), and the above-described LSTM layers for sequential data (i.e., dynamic data). Both branches were then concatenated according to the architecture described in [183], and produced a classification output 'Adherent' or 'Not Adherent'.

For the initial train—test split, we used a temporal 80% training and 20% test split. The temporal split is a common technique in PBPM as it mimics the applied use case where one would use all available training data to predict cases in the future [28]. Therefore, we took all cases that had finished before July 1st, 2022, as training cases, and all cases afterward were considered for the test data set, resulting in an approximate 80:20 split. To ensure that no data leakage occurs by splitting an EC batch into train and test data, the split was conducted on batch level. Furthermore, to similarly ensure that no data spilling occurs between training data and validation data, the training data was likewise randomly split on EC batch level. Additionally, in the same manner, a third of the validation data set was set aside for calibrating the algorithms after training.

For hyperparameter optimisation, we ran a random search over 16 iterations due to its efficiency compared to a full grid search, while still providing equally good results [184]. As the training of the algorithms is not deterministic, each training combination of algorithm, bucket, and parameter set was run 20 times, with the best average performance determining the best parameter set. For final testing, the best parameter set was then applied and tested on the test data set. To ensure comparability across the algorithms, the train—validation—calibration—test split was shared for all algorithms. In total, we trained 30 models: six approaches (as seen in Table 5.7) on five distinct datasets (Table 5.4).

5.4.6 Model Evaluation

To answer the three RQs, specific objective evaluation metrics are required. Regarding RQ1, it can be argued that a classifier is able to predict the outcome, if it performs better than an uninformed trivial classifier [180]. Classifiers are usually tested by comparing different metrics [185]. A common PBPM metric is the *area under the receiver operating characteristics curve* (AUC-ROC) [35], due to it being robust even for imbalanced datasets. As a lower limit value, an uninformed trivial classifier would have an AUC-ROC of 0.5, thus a value much greater than 0.5 indicates a well performing classifier. Furthermore, the *area under the precision recall curve* (AUC-PR) metric is added to provide additional insight into a model's performance when dealing with imbalanced data [186]. Although both are decision threshold independent [186], these metrics have a flaw.

In a business context, a classifier with a higher AUC-ROC or AUC-PR score indicates a better performance only if its convex hull dominates the other classifiers' [187]. In case the ROC curves of two classifiers cross once or more, the question *'Is a classifier better than the other?'* changes to *'When is a classifier better than the other?'* [180]. This can be answered by investigating the ROC graphs [188] and the usage of cost curves [180]. However, this rather time-consuming evaluation procedure is unsuitable for mass evaluation during hyperparameter tuning. Hence, we used a two-stage evaluation principle: First, during hyperparameter tuning, we decided on the best parameter setting based on the AUC-ROC value for each approach, while for final evaluation and answering RQ2, we conducted a visual inspection of the ROC curves, cost curves, and PR curves. While ROC and PR curves are common graphs, cost curves are rarely used. However, they are useful to identify the classifier with the lowest normalised expected cost (NEC) for a given operating condition [180]. The translational formula between ROC- and cost-space is given as [180]:

$$NEC(t, PC(+)) = (FNR(t) - FPR(t)) * PC(+) + FPR(t) \qquad (5.1)$$

with $FNR = 1 - TPR$ being the false negative rate, t representing the decision threshold, and $PC(+) = \frac{c_{FN} * p(+)}{c_{FN} * p(+) + c_{FP} * p(-)}$. The variable $PC(+)$ in this formula denotes the probability times cost, the normalised product of misclassification costs for a false negative c_{FN} and the proportion of positive examples $p(+)$ during deployment [180]. For purposes of generalisation, instead of a predefined absolute value for misclassification the cost proportion c defined as

$$c = \frac{c_{FN}}{c_{FN} + c_{FP}} \tag{5.2}$$

is used. Thus, by substituting $PC(+)$ with the *skew* z [189]

$$z \triangleq \frac{c * p(+)}{c * p(+) + (1-c)(1-p(+))} \tag{5.3}$$

we get:

$$NEC(t, z) = (FNR(t) - FPR(t)) * z + FPR(t) \tag{5.4}$$

For all purposes of evaluation, this formal adaptation does not change the cost curve or its shape, but rather simplifies the discussion. Applying Equation (5.4), a line can be drawn in Cost-space for each point $P_{ROC}(TPR(t), FPR(t))$ in ROC-space (Fig. 5.3). For each point $P_{ROC}(TPR(t), FPR(t))$ the minimal and maximum value of z is inserted into Equation (5.4). Accordingly, the NEC value at $(t, 0)$ is equal to the $FPR(t)$. Correspondingly, the value of $NEC(t, 1) =$

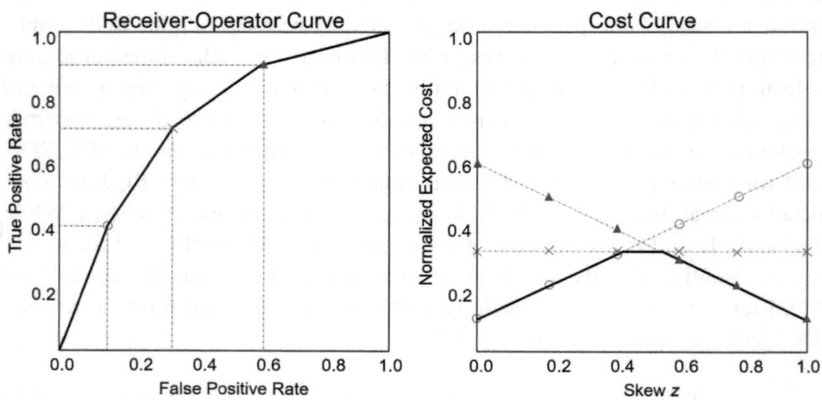

Fig. 5.3 Exemplary transformation from ROC-Space to cost-space for a classifier. Points in ROC-space are straight lines in cost-space and vice versa

$FNR(t)$, which is equal to $1 - FPR(t)$. Connecting the pointwise minimum for each value of z, the lower envelope is constructed. This lower envelope is the cost curve and represents the set of operating conditions (t, z), under which a classifier has the lowest misclassification costs. In case two algorithms are compared, the same logic as in ROC-space applies to cost-space: if one algorithm's cost curve is lower for all values of z than another's, then it dominates [180]. If they cross, it can be directly evaluated which algorithm to use depending on the skew z. Thus, for answering RQ2, we investigate the shape of the respective ROC and cost curves.

For RQ3, we provide a business and domain sensitive discussion of the costs associated with applying ML. This discussion is supplemented by plotting the graphs of (5.4), however, this time we use a predefined value of z over a varying value of t. They are also cost curves, with the limitation that the skew is fixed to the deployment values. We term these curves *deployment cost curves.* To provide a business optimal threshold T_z^o for a given value of z, we take (5.4) and apply the optimal threshold choice method [189]:

$$T_z^o(z) \triangleq \underset{t}{\operatorname{argmin}}\, 2\{(FNR(t) - FPR(t)) * z + FPR(t)\} \tag{5.5}$$

For final interpretation, one more clarification is required: *costs of misclassification.* The above formulas follow the assumption that only costs resulting from ML application are used for calculation [180]. As a result, c_{FN} and c_{FP} are not the full costs of the misclassified case, but rather the difference of costs due to misclassification and correct classification costs. The four base costs are easily defined by asking the following questions: How much will a false positive (FP), a false negative (FN), a true negative (TN), and a true positive (TP) case cost the organisation? We write these base costs as c_{FP0}, c_{FN0}, c_{TN0}, and c_{TP0}. The values for (2) are given as $c_{FN} = c_{FN0} - c_{TP0}$ and $c_{FP} = c_{FP0} - c_{TN0}$. In the EC case, c_{FP} represents the costs incurred from additionally examining specific ECs without necessity. c_{FN} is a combination value. 'Not Adherent' cases that are overlooked cause misclassification costs c_{FN0}. . Correctly classifying 'Not Adherent' cases would however incur costs associated with correcting them towards an 'Adherent' case (c_{TP0}). The difference of these two values is thus not to be interpreted as costs resulting from misclassifying, but rather as the loss of missing out on the chance to correct a 'Not Adherent' case. Classification costs are highly dependent on the business context. Although the industry partner has provided us with approximate values, due to compliance with said partner only the cost proportion

c is reported ($c = 0.8$) and the accompanying relative values of $c_{FN} = 4MU$ and $c_{FP} = 1MU$ (monetary unit). Using these values, as well as the positive proportion for the test sets given in Table 5.4, we can calculate the skew z for each of the data sets.

5.5 Results

Following model development, a demonstration of the artefact shall be given (Step 4 of the DSRM procedure). As our artefact was the application of PBPM for EC effectivity date adherence prediction, we defined the demonstration of predictive performance as the method of choice. Testing the developed models on the split-off test data set can be further seen as the expected performance of the algorithms once deployed as the core of a monitoring tool (shown in Fig. 5.7). All training, hyperparameter optimisation, and final testing was performed on an AWS SageMaker ml.c5.9xlarge instance. All results are displayed in Table 5.7, Fig. 5.4, Fig. 5.5, and Fig. 5.6. Presenting the variance of each curve was omitted due to readability. The thin dashed lines represent the trivial classifiers.

Our results indicate that PBPM methods can be applied for predicting adherence to EC effectivity dates. As seen in Table 5.7, the mean AUC-ROC value of each classifier is above 0.5 for the unified data set, with the best value being highlighted. Furthermore, Table 5.7 shows that the performance gaps of the shallow learning classifiers are neglectable in terms of AUC-ROC. For comparative reason, the F0.5., F1-, F2-scores, and Matthew's correlation coefficient (MCC) for each prediction model at a threshold of 0.5 where added. Finally, the last column displays the lowest NEC at the optimal threshold T_z^o. As shown, none of the other metrics consistently coincides with the lowest NEC, confirming the necessity to introduce cost evaluation apart from AUC-ROC, F-scores, and MCC.

Additionally, Fig. 5.4 and Fig. 5.5 show the ROC, PR, and cost curves for each of the approaches and datasets. The curves represent the mean value by vertical averaging (Fawcett 2004). It is displayed, that within our case-study, none of the approaches is fully dominated by the other, graphically showcasing why it is necessary to evaluate NEC values. Focusing on the cost curves, the results indicate that the best performing classifier for the given data set is only by chance that which is highlighted in Table 5.7. Moreover, if the skew z of the data set and buckets would have been any different, than the best performing classifier might differ.

Table 5.7 Results*

Approach**	AUC-ROC (Mean)	AUC-ROC (Std)	AUC-PR (Mean)	AUC-PR (Std)	F0.5-Score	F1-Score	F2-Score	MCC	Min. NEC	T_z^o
Dataset Full (train = 53 313, test = 9066, skew = 0.536)										
XGB_idx	0.754	0.007	0.470	0.012	0,507	0,536	0,438	0,345	0,303	0,170
XGB_agg	0.750	0.005	0.474	0.007	0,506	0,529	0,446	0,345	0,309	0,231
RF_idx	0.747	0.002	0.508	0.006	0,503	0,548	0,405	0,333	0,294	0,248
RF_agg	0.758	0.004	0.491	0.010	0,518	0,614	0,352	0,340	0,301	0,199
BNN	0.714	0.011	0.436	0.015	0,474	0,497	0,421	0,308	0,337	0,242
LSTM	0.715	0.010	0.431	0.019	0,459	0,463	0,448	0,300	0,336	0,214
Dataset 0.0 (train = 30 793, test = 6085, skew = 0.547)										
XGB_idx	0.747	0.006	0.478	0.017	0,524	0,536	0,493	0,370	0,307	0,261
XGB_agg	0.767	0.005	0.480	0.009	0,516	0,531	0,477	0,358	0,289	0,203
RF_idx	0.752	0.001	0.534	0.006	0,548	0,581	0,467	0,385	0,291	0,341
RF_agg	0.756	0.002	0.501	0.005	0,604	0,683	0,447	0,431	0,296	0,190
BNN	0.729	0.012	0.453	0.020	0,491	0,523	0,420	0,318	0,322	0,240
LSTM	0.688	0.019	0.404	0.030	0,405	0,395	0,441	0,232	0,348	0,093
Dataset 1.0 (train = 16 226, test = 2529, skew = 0.435)										
XGB_idx	0.782	0.012	0.450	0.020	0,484	0,534	0,379	0,357	0,287	0,238
XGB_agg	0.741	0.005	0.527	0.007	0,515	0,548	0,437	0,393	0,285	0,407
RF_idx	0.723	0.007	0.420	0.012	0,412	0,442	0,343	0,279	0,323	0,236
RF_agg	0.778	0.004	0.553	0.004	0,598	0,741	0,380	0,441	0,285	0,403
BNN	0.697	0.018	0.350	0.023	0,391	0,482	0,253	0,260	0,339	0,255
LSTM	0.690	0.023	0.327	0.030	0,328	0,312	0,388	0,219	0,345	0,372

(continued)

Table 5.7 (continued)

Approach**	AUC-ROC (Mean)	AUC-ROC (Std)	AUC-PR (Mean)	AUC-PR (Std)	F0.5-Score	F1-Score	F2-Score	MCC	Min. NEC	T_z^o
Dataset 0.1 (train = 3306, test = 345, skew = 0.765)										
XGB_idx	0.428	0.100	0.529	0.075	0,351	0,519	0,212	−0,022	0,207	0,109
XGB_agg	0.337	0.055	0.457	0.032	0,339	0,441	0,210	−0,119	0,224	0,066
RF_idx	0.356	0.094	0.460	0.045	0,097	0,217	0,036	−0,149	0,208	0,114
RF_agg	0.441	0.034	0.505	0.022	0,118	0,260	0,045	−0,155	0,210	0,159
BNN	0.323	0.102	0.482	0.075	0,234	0,575	0,168	−0,020	0,317	0,064
LSTM	0.365	0.148	0.494	0.098	0,465	0,451	0,535	−0,222	0,230	0,227
Dataset 1.1 (train = 2988, test = 107, skew = 0.479)										
XGB_idx	0.830	0.028	0.459	0.046	0,428	0,519	0,283	0,215	0,213	0,163
XGB_agg	0.800	0.017	0.585	0.021	0,500	0,487	0,560	0,362	0,259	0,377
RF_idx	0.831	0.006	0.604	0.029	0,532	0,583	0,422	0,397	0,219	0,297
RF_agg	0.751	0.020	0.424	0.022	0,594	0,626	0,520	0,447	0,232	0,433
BNN	0.521	0.048	0.214	0.035	0,186	0,229	0,155	−0,014	0,414	0,251
LSTM	0.461	0.091	0.180	0.050	0,146	0,131	0,223	−0,006	0,427	0,549

*best value for metric and bucket highlighted in **bold**

**Approach:	Learning Algorithm:	Encoding Technique:
XGB_idx	XGBoost	Index
XGB_agg	XGBoost	Aggregation
RF_idx	Random Forest	Index
RF_agg	Random Forest	Aggregation
BNN	Branching NN	Index
LSTM	LSTM NN	Index

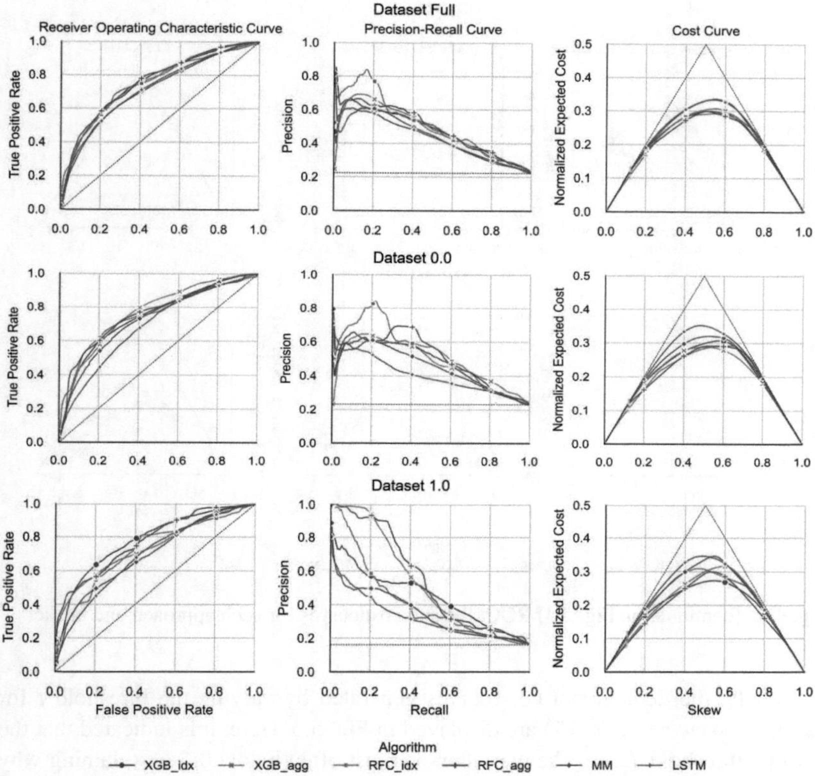

Fig. 5.4 ROC, PR and cost curves for each data set and approach

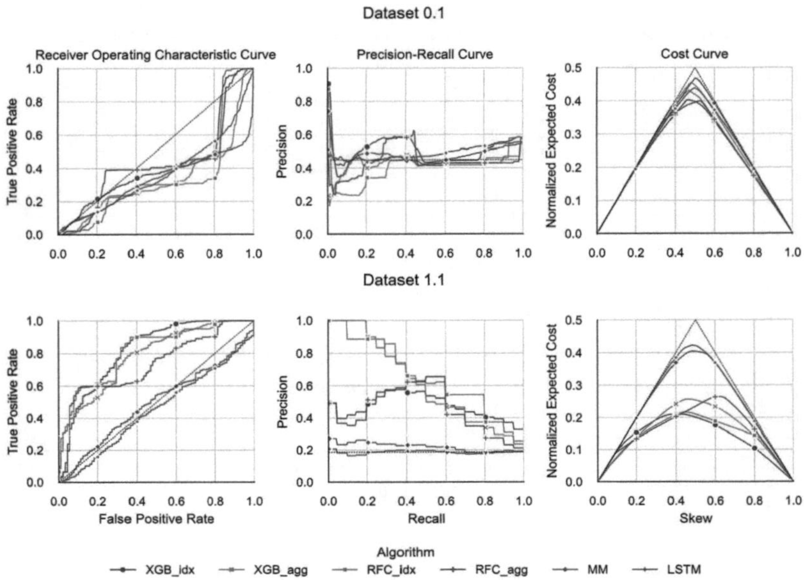

Fig. 5.5 (continuation Fig. 5.4) ROC, PR and cost curves for each approach and bucket

Finally, the deployment cost curves generated by varying the threshold t for the given skew z (Table 5.7) are displayed in Fig. 5.6. Here, it is indicated that the optimal threshold T_z^o for the classifiers is most often below 0.5—explaining why F-scores and MCC values for 0.5 do not indicate the best performing classifier. It is further shown how the NEC vary based on the threshold, and why detailed investigation of cost curves and deployment cost curves is necessary to decide on which classifier to operationalise.

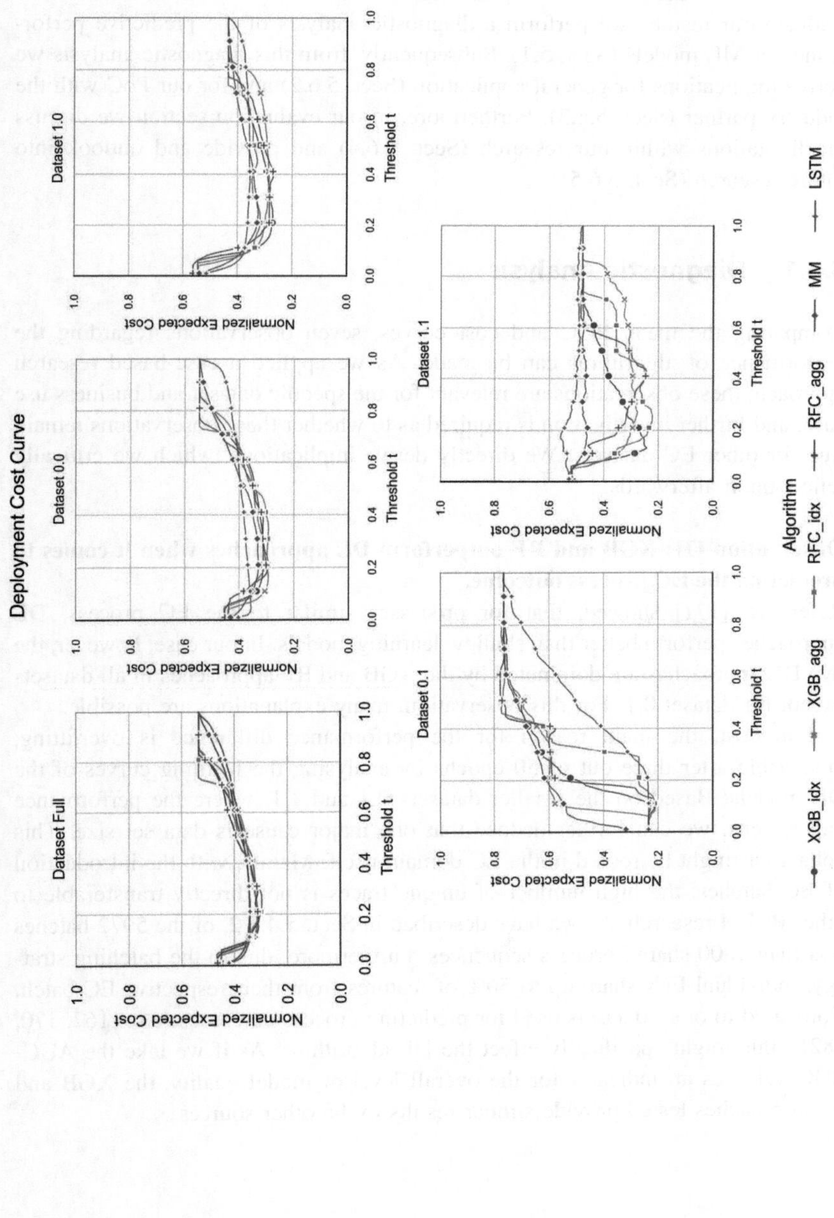

Fig. 5.6 Deployment cost curves for each approach and bucket

5.6 Discussion

The second to last step according to DSRM is the evaluation of the artefact. To evaluate our results, we perform a diagnostic analysis of the predictive performance of ML models (Sect. 5.1). Subsequently, from this diagnostic analysis we derive implications for general application (Sect. 5.6.2) and for our PoC with the industry partner (Sect. 5.6.3). Furthermore, in our evaluation section we discuss the limitations within our research (Sect. 5.6.4) and provide and outlook into future research (Sect. 5.6.5).

5.6.1 Diagnostic Analysis

Comparing the mean ROC and cost curves, seven observations regarding the performance of algorithms can be made. As we applied a case-based research approach, these observations are relevant for the specific dataset and business use case, and further investigation is required as to whether these observations remain true for other EC datasets. We directly derive implications, which we critically reflect upon afterwards.

Observation O1: XGB and RF outperform DL approaches when it comes to predicting the EC process outcome.
Reference [171] showed, that for processes similar to the EC process, DL approaches perform better than shallow learning models. In our case, however, the two DL approaches are dominated by the XGB and RF approaches in all datasets except for dataset 0.1. For this observation, many explanations are possible.

Foremost, the main reason for the performance difference is overfitting, observable after three out of 50 epochs by analysing the learning curves of the DL models. Based on the smaller datasets 0.1 and 1.1, where the performance gap widens, we can further deduce that one major cause is data set size. This behaviour might be rooted in the EC domain itself. Mainly, with the introduction of EC batches, the high number of unique traces is not directly transferable to other PBPM research. As we have described in Sect. 5.4.1.2, of the 5972 batches less than 1000 shared process sequences. Furthermore, due to the batching strategy, individual ECs share up to 50% of features from their respective EC batch. Compared to other datasets used for predicting process outcomes (e.g. [167, 170, 182]), this might specifically affect the DL algorithms. As if we take the AUC-ROC value as an indicator for the overall level of model quality, the XGB and RF approaches tested provide similar results to the other sources.

Another contributor to the worse performance of DL algorithms is the algorithm and data structure. McElfresh et al. [190] have shown that DL algorithms, especially NN, perform worse than bagging and boosting algorithms (RF and XGB respectively) when it comes to tabular data. Specifically for larger data sets with a high number of features, the NN classifiers perform worse. These characteristics are also available with our data set: There are relatively many instances (>50 000) with high dimensionality (>1000 for index-based encoding). This high dimensionality introduces noise to the data, which can be easier processed by shallow learners [191]. Moreover, LSTM networks and the BNN were designed for text data, wherein one major effect running against the NNs' strength lies: DL algorithms are known to excel shallow learners when it comes to unstructured data. The tabular data used in our case study, however, already prepares the features for direct consumption by the shallow learners.

Lastly, the effects of algorithm architecture of the DL algorithms can be neglected. The LSTM algorithm applied is a revised and tested architecture already employed in PBPM literature (e.g. [167, 170]), where it achieved comparable results to XGB, and the BNN approach performed similarly.

Implication I1: Compared to other PBPM use-cases, DL for predicting EC effectivity date adherence is not suitable

O2: There is no clear indication whether XGB-based approaches outperform RF.

Within our experiments, another unexpected result was observed. In general ML literature, XGB approaches often perform better than those based on RF. This observation could not be confirmed by our experiments. At first glance, it would be indicated that XGB-based approaches benefit the most from bucketing, as they have consistently produced the lowest NEC values. However, if we consider the cost curves, this is sheer coincidence due to the skew of the test set. Most cost curves show that for a lower value of skew, an RF based approach would perform better than XGB. Similarly, both RF and XGB based approaches had at least one bucket with the highest AUC-ROC.

This observed performance difference can be caused by many factors. A major factor within our research was the application of the calibration run. Following this, the average AUC-ROC for RF_agg increased from 0.736 to 0.758, whereas the AUC-ROC for XGB_agg, XGB_idx, and RF_agg dropped from 0.772, 0.773, and 0.763 to 0.754, 0.750, and 0.747, thereby changing the ranking of the algorithms. Screening the respective literature, except for [170], none of the other publications mentions a calibration run.

Furthermore, examining the learning curves of the runs, we were able to identify overfitting of the XGB algorithm after 100 epochs, while the RF classifier increases accuracy until the end, which could be another cause of the comparably lower performance of the XGB. This again is rooted in the algorithms' inner workings, as RF are known to be more resistant to overfitting. The common countermeasure to such overfitting would be providing more data.

I2: Calibrating the trained ML models can have a significant impact on predictive performance

I3: Predicting EC effectivity date adherence with XGB requires more data than comparable PBPM use-cases.

O3: No preference towards aggregation-based encoding or index-based encoding can be shown based on predictive performance.
As can be seen in the graphs and mean values, neither the index nor the aggregation-based approaches dominate the other. In fact, each algorithm and encoding combination has once performed best regarding AUC-ROC. The second evaluation metric, expected cost, does also not provide an indication, as depending on the skew, different approaches and algorithms show better performance. As both encoding approaches performed better than a no-skill classifier, a hybridisation via for instance majority vote classifier could increase classification performance.

I4: As index-based and aggregation-based encoding perform similarly, no preference regarding an encoding technique can be given.

O4: Bucketing has no evident impact on relative model quality.
For all metrics, there are some algorithms benefiting from bucketing, while others are impaired by it. Even for the seemingly best performing bucket, dataset 1.1, although XGB_idx, XGB_agg and RF_idx increase their AUC-ROC score, RF_agg has a lower AUC-ROC score (the NNs are affected negatively as well due to less data availability, see O1). This change in predictive performance is possibly rooted in the reduction of training instances.

I5: Bucketing should be considered as a method to improve model quality, but a comparison against the full dataset is required.

O5: The schedule adherence of quality triggered EC cases is not determined by EC data.
ROC, PR and cost curve graphs for dataset 0.1. stand out among those shown in Fig. 5.4 and Fig. 5.5. It is for this dataset alone that all algorithms perform worse than a no-skill algorithm. As dataset 1.1. has less cases, the reduced performance cannot be fully attributed to the reduced number of traces. Moreover, the worse results indicate that for this bucket wrong features are used, i.e., other process and case characteristics not used in the data model determine the process outcome. This could be a result of focusing on data available from the main EC process path and omitting data from supporting processes such as the commercial approval process and total quality control.

I6: Bucketing based on domain knowledge could help reveal bucket-specific deficiencies in the data model.

O6: In an applied business context, a higher AUC-ROC does not necessarily point to the better algorithm.
As seen in Table 5.7, only for Dataset 0.0 did the best performing algorithm regarding AUC-ROC also have the lowest NEC (XGB_idx). For all other cases, this was not observable. This can also be seen in the respective cost curves. In fact, none of the other metrics had an observable correlation with the NEC, demonstrating that in a business context, the expected deployment skew is paramount for choosing the best classifier.

I7: In a real business deployment scenario, NEC values are the best metric to choose the best performing classifier.

O7: A low NEC is not necessarily the result of a good classifier.
In O4, we did not discuss the effects of bucketing on the NEC value, as it cannot be trivially confirmed. By analysing the cost curves, we can see that the maximum possible NEC value for a classifier is determined by the skew. This effect is most drastically seen in dataset 0.1. Of all buckets, it has the lowest NEC. However, comparing the deployment cost curves for dataset 0.1, all approaches' deployment cost curves are steadily increasing from the lowest cost of approx. 0.2. Investigating this curve and comparing values with the other datasets, this

relates to the skew. Dataset 0.1 has a nearly balanced class ratio in the test set (see Table 5.4). As the c_{FN} outweighs c_{FP} 4:1, the resulting skew is ~ 0.75. Thus, a trivial positive classifier always predicting class 1 and double-checking every case would result in 189,75MU for needless doublechecking vs. savings of 621MU, or a NEC of 0.23 which is only slightly worse than the best classifier. As a result, it is not sufficient to say that a NEC below 0.5 equals a good classifier, but it is imperative that the classifier performs significantly better than the trivial classifiers for the given skew.

I8: For heavily skewed data the tested models are not sufficient to justify expensive ML solutions.
In total, we were able to show that applying PBPM for EC effectivity date adherence prediction is possible. However, compared to other research within the domain, the shallow learners outperformed DL algorithms. Although the BNN sometimes slightly outperforms the LSTM network, it performs worse than the XGB and RF classifiers. The overall worse performance of NNs and similar performance of XGB and RF indicate overfitting, which could be due to scarcity of data (I1 and I3).

5.6.2 Implications for the Prediction of Effectivity Date Adherence

From the results, we can state that effectivity date adherence prediction is possible three weeks before first delivery. Our analysis shows that shallow learning algorithms outperform DL algorithms, given sufficient feature engineering and data (I1 and I3).

The results further show that for most buckets and skews, shallow learning classifiers perform better than trivial classifiers in predicting the effectivity date adherence, answering RQ1: ML application can be utilised to predict the adherence to an EC effectivity date. Different future research paths could also enhance performance. From I3 we can derive that there is potential to improve encoding by combining aggregation-based and index-based encoding techniques. For instance, time related features of specific milestones, e.g., '*purchase order confirmation*' or '*EC approval*', could be extracted and added to aggregation-based encoding. From I5 and I6 we can derive that bucketing is not always necessary, but it can help to better tailor the data model to the respective process.

Taking into consideration I1, I3, I5, and I7 regarding RQ2 it can be said that there is no clear indication which approach is best, as no single approach dominates all others in ROC- or cost-space. Hence, for answering RQ2, we refer to the cost curves, to decide based on an economical consideration which algorithm has the best NEC for the specific skew. Additionally, predictive performance could be quickly improved by a majority vote classification with only little effort.

5.6.3 Functional Testing and Deployment

Following the theoretical analysis, a PoC tool was developed for monitoring the EC effectivity date adherence and its prediction for cases not yet implemented. Figure 5.7 displays the frontend of the tool, tested at the industry partner. In the backend, a decision algorithm is integrated for choosing the best ML approach, derived from the deployment and application phase shown in Fig. 5.8. For that, the past month's deployment skew and the ML algorithms cost curves are taken for reference. From the discussions with the industry partner, one major comment was raised regarding trust in AI:

Management was sincerely interested in understanding the models' predictive performance and the metrics. They remarked, that the models are far off from unrestricted applicability in a productive environment. To make that possible, the classifiers' performance would need to improve to ensure acceptance by operating staff. As a guideline, it was noted that for unrestricted applicability and acceptance by the staff, the ratio of TP:FP should be 9:1.

However, this value was rarely achieved by the algorithms, with the best having a precision of ~ 0.8 at an acceptable recall value. Management further remarked that during implementation of ML models, it would be more important to ensure employee acceptance, rather than operate at the cost optimal threshold T_z^o. Therefore, instead of presenting the employees with the predicted class, the probability of 'Adherent' or 'Not Adherent' would be given, effectively presenting the organisation with a ranking system (seen in the last column in Fig. 5.7). For those cases that have a predicted failure probability of $p^+ > 0.9$, all would be considered, and for those with a value $0.5 < p^+ \leq 0.9$, only critical changes with high monetary value would be considered.

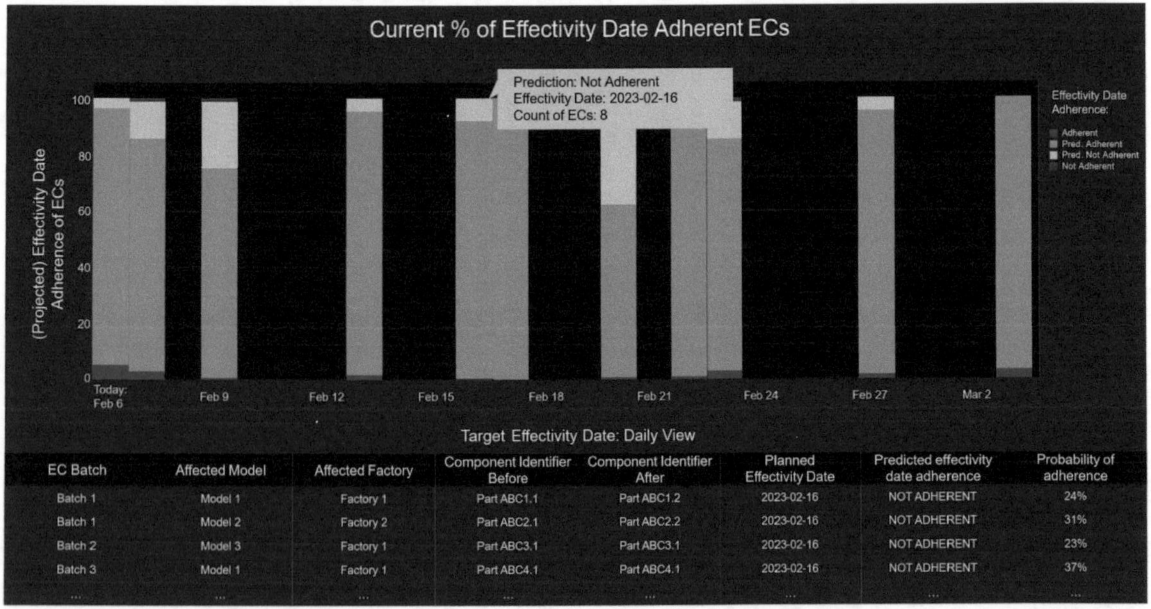

EC Batch	Affected Model	Affected Factory	Component Identifier Before	Component Identifier After	Planned Effectivity Date	Predicted effectivity date adherence	Probability of adherence
Batch 1	Model 1	Factory 1	Part ABC1.1	Part ABC1.2	2023-02-16	NOT ADHERENT	24%
Batch 1	Model 2	Factory 2	Part ABC2.1	Part ABC2.2	2023-02-16	NOT ADHERENT	31%
Batch 2	Model 3	Factory 1	Part ABC3.1	Part ABC3.1	2023-02-16	NOT ADHERENT	23%
Batch 3	Model 1	Factory 1	Part ABC4.1	Part ABC4.1	2023-02-16	NOT ADHERENT	37%
...	...	...	...	...	...	...	...

Fig. 5.7 Screenshot of the proof of concept for monitoring the actual and predicted EC effectivity date. Due to confidential information, the EC effectivity date and EC specific information was replaced with dummy values. The x-axis shows each day within the monitoring period of three weeks. The bar graph shows the proportion of “Adherent”, “Predicted Adherent”, “Predicted Not Adherent”, and “Not Adherent” ECs scheduled for that date. The below table shows detailed information based on the selected date and adherence projection, in this case February 16th, “Predicted Not Adherent”

In conclusion, and as an answer to RQ3 we can note that from a purely economic point of view, it is sensible to apply ML to predict effectivity date adherence. This should always be done while taking consideration of the actual skew (I6), as for some extreme skews, it is more beneficial to double check everything instead of maintaining an ML model.

5.6.4 Limitations

Having discussed the results and implications of our research, several limitations require mentioning. Due to restricted data availability, we considered only data from the EC process of a single OEM, which could have its own peculiarities, as the EC process is tailored to individual companies and their current requirements. Consequently, although in our research case bucketing did not lead to significant improvements, for other use-cases this may very well be the case. Furthermore, other industries and companies might have different features available. Likewise, ECs are not a rare phenomenon in the automotive industry. As such, associates are experienced and the EC process fleshed out. Other effects stemming from peculiarities of the automotive industry, e.g., products and concepts often being reiterations cannot be determined without comparative studies in other industries.

Here, the availability and validity of data can pose a main challenge. For good predictive performance, a thorough replication of reality must be available, which we have encountered at the OEM. Thus, the base data quality was already sufficient to perform ML operations. Furthermore, as seen with the overfitting XGBoost and DL algorithms, a lot of case data is necessary. In organisations which have fewer ECs, this might not be the case. Increasing the time frame of the data collection might also not be ideal, as processes change over time, and some aspects which had been relevant in cases long ago might not be relevant today. Regardless of these limitations, we were able to answer all RQs and demonstrated how ML and PBPM can be applied in the EC domain for schedule prediction.

5.6.5 Fulfilment of Objectives and Future Work

Ultimately, we can state that our artefact fulfilled the three objectives stated in Sect. 5.3.3: Regarding objective 1, it was shown, that ML and PBPM can be applied to predict the adherence to a given schedule. To fulfil objective 2, it was demonstrated that there is no single general "best approach", but rather a case-by-case answer depending on the deployment skew. Targeting objective 3, the business-oriented evaluation metric, we have shown how to apply cost curves for model evaluation. As part of our validation with the industry partner, we have furthermore shown that when implementing and applying new technology in a productive environment, other factors besides economic values come into play. It becomes necessary to ensure employee and management acceptance. Therefore, in a productive environment, the targets of ML application should be clarified first, before deciding on the operating conditions.

With our research, we have provided a blueprint of how to determine which algorithm performs best from a business perspective given a specific deployment condition. The resulting procedure and the PoC are graphically shown in Fig. 5.7 and Fig. 5.8.

From this baseline, several future research paths can be taken. First, we have applied the approaches at the end of the implementation phase, three weeks before the EC effectivity date. However, within this timeframe, only limited counteraction can be undertaken. Thus, to improve the monitoring method, it would be necessary to understand how early ML and PBPM can be applied to predict the adherence to the schedule. Secondly, the ML algorithms applied and features selected should be further developed. In our research, we focused exclusively on ECM data. However, within an organisation, testing other data systems, e.g. procurement and logistics, could show whether they improve predictive performance. Lastly, the transfer of the results into other industries should be investigated. From literature, it can be deduced that mass customisation industries are facing other challenges in EC than engineer-to-order companies or mass production.

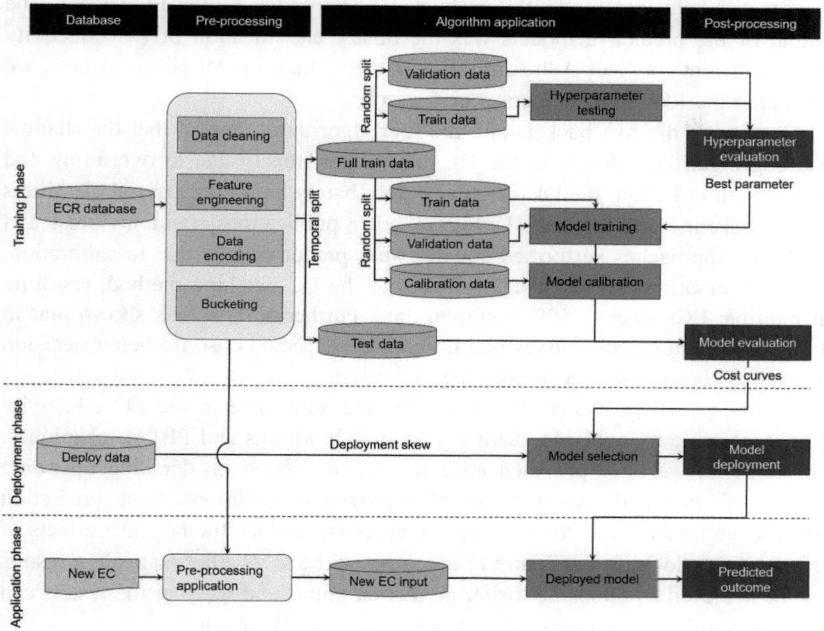

Fig. 5.8 Suggested workflow for training, deployment, and application of machine learning models for ECM

5.7 Conclusion

Changing market demands, design errors, and new technologies require frequent changes to products. With increasing complexity of products, it is crucial that co-dependent ECs are implemented simultaneously. If not, deviations from the effectivity date can cause increased cost of quality or product malfunction. To support associates with guidance on what to focus on, we trained and evaluated a total of 30 ML-models, which differ in applied learning algorithm, data encoding technique, and EC type. For training and evaluation, we used traces from an event log provided by a German automotive OEM.

To provide a wide variety of approaches, four algorithms were tested: Random Forest and XGBoost as representatives of shallow learning algorithms, and an LSTM neural net and a branching neural net as DL approaches. The event log was prepared by means of aggregation and index-based encoding. Afterwards, the dataset was split into four different buckets, determined by the EC trigger. The output of the predictive models was the binary classification of EC effectivity date 'Adherent' or 'Not Adherent'. To evaluate which model performs best, we compared the ROC, PR, and cost curve graphs.

Comparing the five buckets and the four algorithms showed that the shallow learning algorithms dominate the DL approaches mostly due to overfitting and data structure. In fact, the DL classifiers were barely better than trivial classifiers when bucketing was applied. Unlike in other publications, random forest and XGBoost approaches performed equally well, predominantly due to calibration. These observations are most likely linked to the EC batching method, resulting in multiple ECs sharing 50% common data. Furthermore, it was shown that in a business context, cost curves and normalised-expected-cost are better decision metrics to choose the best performing algorithm.

Overall, we have shown that predicting the adherence to the EC effectivity date is possible by applying shallow learning algorithms and PBPM techniques. Furthermore, we have provided an extensive discussion on deciding operating points and thresholds based on actual deployment conditions. Such prediction models can be useful to steer scarce resources and reduce the negative effects of EC. As an outlook, in our future research we will test whether equivalent results can be achieved by choosing earlier prediction points and whether the results can be improved by incorporating data from supporting processes.

5.8 Presentation of Own Contribution

All data presented within the publication was collected and the contents processed by me. This includes the definition of the research scope and research questions, screening of theoretical background research, identification of relevant data bases, definition of test sets, identification and application of the learning algorithms, and definition of the evaluation criteria. Furthermore, the implementation of the learning algorithms into Python code, the execution of the test code, and the evaluation of the results was done by me. Additionally, the textual elaboration and writing of the publication were entirely conducted by me. All four co-authors Prof. Dr. Burggräf, Dr. Steinberg, Mr. Becher, and Dr. Weisser contributed with ideas to the research concept and in the internal review process.

Publication IV: Evaluating Early Predictive Performance of Machine Learning Approaches for Engineering Change Schedule—A Case Study Using Predictive Process Monitoring Techniques

6

6.1 Abstract

By applying machine learning algorithms, predictive business process monitoring (PBPM) techniques provide an opportunity to counteract undesired outcomes of processes. An especially complex variation of business processes is the engineering change (EC) process. Here, failing to adhere to planned implementation dates can have severe impacts on assembly lines, and it is paramount that potential negative cases are identified as early as possible. Current PBPM research, however, has seldomly investigated the predictive performance of machine learning approaches and their applicability at early process steps, let alone for the EC process. In our research, we show that given adequate feature encoding, shallow learners can accurately predict schedule adherence after process initialisation. Based on EC data from an automotive manufacturer, we provide a case sensitive performance overview on algorithm-encoding combinations. For that, three algorithms (XGBoost, Random Forest, LSTM) were combined with four encoding techniques. The encoding techniques used were the two common aggregation-based and index-based last state encoding, and two new combinations of these, which we term advanced aggregation-based and complex aggregation-based encoding. The study indicates that XGBoost index encoded

---

Radisic-Aberger, O., Burggräf, P., Steinberg, F., Becher, A., & Weisser, T. (2024). Evaluating early predictive performance of machine learning approaches for engineering change schedule—A case study using predictive process monitoring techniques. Supply Chain Analytics, Volume 8, 100087 https://doi.org/10.1016/j.sca.2024.100087.

O. Radišić-Aberger, *Optimisation and Control of Engineering Change Schedules in the Automotive Industry with Metaheuristics and Machine Learning*, Findings from Production Management Research,
https://doi.org/10.1007/978-3-658-51074-9_6

approaches outclass regarding average predictive performance, whereas Random Forest aggregation encoded approaches perform better regarding temporal stability due to reduced influence by dynamic features. Our research provides a case-based reasoning approach for deciding on which algorithm-encoding combination and evaluation metrics to apply. In doing so, we provide a blueprint for an early warning and monitoring method within the EC process and other similarly complex processes.

6.2 Introduction

Modifications of components occur frequently during the lifecycle of a complex product [16] causing various disruptions in running production systems [44]. These changes to the function, behaviour, or structure of a component are known as engineering changes (EC) [4]. From a product perspective, some ECs can be subtle changes to a single component or have modify entire sub-systems due to the phenomenon known as change propagation [13]. Consequently, the administrative strain on the production control system varies: While engineering departments solve design issues, purchasing departments renegotiate with suppliers, and material is steered by supply chain logistics [14, 164]. As a result, ECs and the underlying EC process are complex by nature [1]. For easier management, change coordination boards (CCB) define deadlines for activities, the final being the target EC implementation date known as the *EC effectivity date* [24, 55, 164]. However, regardless of recent advances in process mining and artificial intelligence (AI), monitoring EC process adherence regarding deadlines, especially during their implementation, remains a manual task [44].

With thousands of ECs running in parallel at any given time, this is a dissatisfying state [164]. Even with extensive planning, challenges such as hidden change propagation [192], late tool changes, or late solution adaptations can arise [193]. Subsequently, ECs cause inaccuracies in supply and demand putting further strain on limited materials planning resources [17]. Ranging from lost profits due to scrapping obsolete components and expedited transports to halting the assembly line if components are not available, the economic effects of late (or early) ECs can have a significant impact on a company's future [165]. In this complex environment, EC coordinators and CCBs face a dilemma: How to distinguish ECs running off-schedule from ones which can be processed without further supervision with limited administrative capacity?

In support of this challenge, it is suggested to investigate whether automation via machine learning (ML) could be applied for early monitoring and prediction of EC scheduling adherence [1, 44]. The EC specific task can be generalised as 'What is the minimum process length to predict and monitor the outcome of a complex process'. It lies at the core of predictive business process monitoring (PBPM), a technique set from the research fields of process mining and ML [169]. Researchers have shown that it is possible to predict the availability of components at assembly lines [168] or whether a patient will be treated on time [35]. Most publications have predicted the outcome of the process at predefined moments, for example at specific process steps [194] or as an offset to a deadline ([35, 167, 171, 195]). Regardless, all features seen during ML model training are available at the time of prediction. However, early prediction leads to a situation where features available at later process steps might not be available yet (e.g. feedback from suppliers and contract information), and the outcome is case and domain sensitive [170]. Therefore, to advance the possibilities of computational support and further leverage the digital thread in engineering change management (ECM) [20], an investigation on early application of ML-based approaches for EC effectivity date adherence prediction is required.

To close this research gap, in this publication we present the outcome of a case study based on EC data from an automotive manufacturer targeted at understanding *how early can machine learning be applied to predict and monitor the adherence to EC effectivity dates?* To answer this overarching question, four comprehensive research questions (RQ) are raised:

- RQ1: How is predictive performance impacted by EC process length?
- RQ2: How is predictive performance impacted by EC case length?
- RQ3: Which algorithm and encoding combination provides the most temporally stable prediction score?
- RQ4: How does process data influence the prediction outcome?

RQ1 and RQ2 are extended questions targeted at understanding the predictive performance of various classifiers over the course of a process. To differentiate, whereas process length is synonymous with *prefix length* (cf. sect. 6.3.2), case length is the total recorded length of the finished case. Generally, predictive performance, measured as area-under-the-curve of the receiver-operator-characteristic (AUC-ROC) [185, 188], is expected to increase with added data. RQ3 targets at a specific PBPM problem: There is little benefit in a classifiers' prediction if the predicted outcome changes with each step. To evaluate a classifier's prediction stability as input data is modified, TEINEMAA ET AL. [170] introduced the concept of 'temporal stability' (TS). Lastly, RQ4 is aimed at understanding whether data modifications in later process steps have an actual impact on the prediction. This is necessary, since if the models cannot draw new knowledge from dynamic data, then there is no advantage to continuous monitoring of effectivity dates.

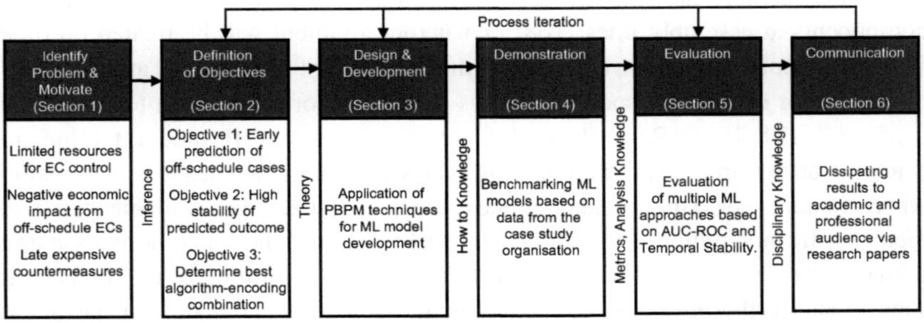

Fig. 6.1 Applied DSRM methodology according to [172] and respective sections in our publication

For guidance, due to the applied interdisciplinary research area of ECM, supply chain management, and information systems, we employ the design science research methodology (DSRM) [172] as our overarching research methodology (Fig. 6.1) As such, this publication classifies as an 'exaptation' (application of a known solution to a new domain) and provides a 'Level 2' artefact (showcasing a method) [173] within the wider Design Science Research (DSR) area. In doing so, we provide two main research contributions, adding to the ECM goals [12] of 'earlier', 'more efficient', and 'more effective': First, we show that it is possible to predict the adherence to planned EC effectivity dates well in advance. Second, we provide a blueprint for benchmarking early application of ML in ECM and similarly complex business processes.

The remainder of our publication is structured as follows: Section 6.3 provides an overall background on EC and PBPM and provides more detail to the RQs, whereas Sect. 6.4 details the methods and resources used in our study. Section 6.5 presents the results of the benchmark, structured by RQ answered, which are discussed in detail in Sect. 6.6. Finally, a summary is given in Sect. 6.7.

6.3 Background

6.3.1 Challenges within Engineering Change Processes

As introduced, managing ECs is demanding. A main source of difficulties is the complexity inherent to the process itself. Although generic EC process descriptions have been established, e.g. [43], the observable instances in productive environments are more circumstantial and company specific [30, 31]. It is seldomly executed according to a generic process flow with frequent loops and parallel processing. As ECs are sometimes applied in critical situations with ad-hoc actions [63], the order of process steps is not definitive. Even more so, all steps are optional and case specific. In such cases, it can occur that the EC is implemented before it is formally approved. Hence, apart from the starting point, the EC trigger, the only other definitive point in the EC process is its implementation on the *EC effectivity date*. It signals the endpoint at which an EC has been realised, and post-change components are available [55].

Another source of EC complexity is the phenomenon of change propagation. Several tools have already been envisioned to help identify propagation paths (see [16] for a thorough review), however, one main mitigation strategy has effects on EC implementation planning: *EC batching*. It was suggested by literature to introduce ECs in groups to reduce interface complexity [15, 63]. As a side effect, this helps CCBs identify codependent components which must be introduced concurrently at the assembly line to ensure buildability. However, if not all ECs of an EC batch are introduced simultaneously, in the worst case, the assembly line must be stopped, costing the production company millions [164].

Thus, it is necessary for CCBs to have a good overview over currently running ECs to act in case of delays. Due to the complex nature of EC, this is not an easy task, and computational aid has been suggested.

6.3.2 Predictive Business Process Monitoring

The overall aim of PBPM application is the prediction of a categorical or numerical process outcome. To achieve this, PBPM follows a generic workflow, shown in Fig. 6.2.

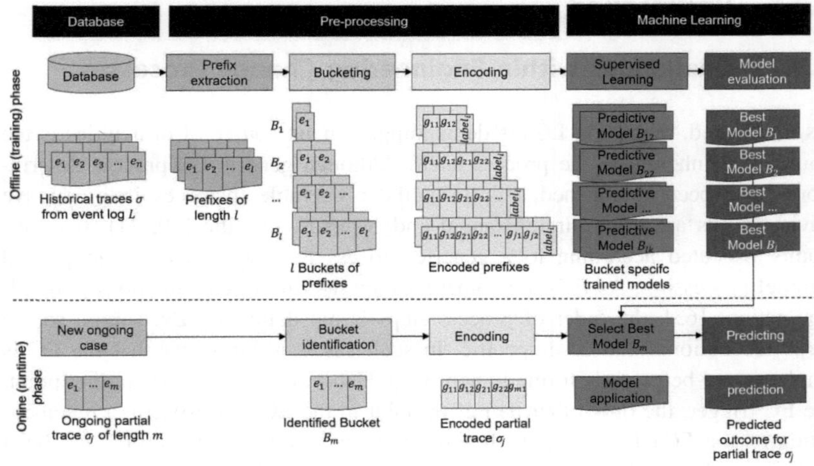

Fig. 6.2 Typical workflow for PBPM application with prefix-length based bucketing, adapted from [28]

For categorical applications, the overall goal is to train classifiers to estimate a class label based on data gathered from an *event log L* of finished *cases,* [28]. Table 6.1 gives an example of such an event log of an EC process, which is further used as a running example.

Each case consists of a collection of unique *events e*, which are documented executions of an *activity* at a recorded *time*, referenced by matching *case IDs*. Additionally, data is recorded along with the events. If modified due to the process execution it is referred to as *dynamic data.* Opposed to that, case data or *static data* is readily available from the beginning of the process, and changes hardly ever [28]. Formally:

- **Definition 1 (Event):** An *event* $e \in \mathcal{E}$ is a tuple $e = (a, c, t, (d_1, v_1), \ldots, (d_m, v_m))$, where a is the name of an executed activity, c the case id, t the timestamp and $(d_1, v_1), \ldots, (d_m, v_m)$ the associated static or dynamic data. $\mathcal{E}$ denotes the universe of events.

The full sequence of events for a given case form a trace σ. Formally:

- **Definition 2 (Trace):** A trace is a non-empty sequence of events $\sigma = [e_1, \ldots, e_n]$ which all refer to the same case.
- **Definition 3 (Case Length):** The full length n of a trace $\sigma = [e_1, \ldots, e_n]$.

Ingesting event log data, learning algorithms are then trained for predicting the outcome of ongoing cases. However, to simulate unfinished cases during the training phase, traces are truncated to shorter *prefixes* [35]. Formally:

- **Definition 4 (Prefix):** For any given trace $\sigma = [e_1, \ldots, e_n]$ of length n, a *prefix*$(\sigma, l) = [e_1, \ldots, e_l]$ can be generated as long as $l \leq n$.
- **Definition 5 (Process Length):** The length l of a *prefix*$(\sigma, l) = [e_1, \ldots, e_l]$. Note, that *prefixes* can assume any length between 1 and $l \leq n$.

Thus, to clarify for RQ1 and RQ2: a finished trace of n events, has the case length n, and can be truncated to prefixes of process length 1 to n. In the case of EC 1 in Table 6.1, this would result in a case length of 11, and prefixes of process length 1, 2, ..., 11.

Table 6.1 An example event log for an EC

Case Info:	Affected Model	Affected Plant		Affected Component
EC 1	M1, M2	P1		C1, C2, C3
Case ID	Activity	Actual Timestamp (DD/MM)	Target Date (DD/MM)	Payload
EC 1	EC triggered	01 / 09	–	–
EC 1	Create new EC request	02 / 09	–	–
EC 1	Calculate commercial impact	10 / 09	–	Cost beneficial
EC 1	Approval	10 / 09	12 / 09	–
EC 1	Release new design	11 / 09	13 / 09	–

(continued)

Table 6.1 (continued)

Case Info:	Affected Model	Affected Plant		Affected Component
EC 1	M1, M2	P1		C1, C2, C3
Case ID	Activity	Actual Timestamp (DD/MM)	Target Date (DD/MM)	Payload
EC 1	Make new tools	30 / 09	–	–
EC 1	Test components	05 / 10	–	Component Not OK
EC 1	Make new tools	20 / 10	–	–
EC 1	Test components	21 / 10	–	Component OK
EC 1	Issue purchase order	22 / 10	14 / 10	–
EC 1	Order new components	22 / 10	17 / 10	–
EC 1	Components delivered	30 / 10	22 / 10	–
EC 1	Implement EC	31 / 10	24 / 10	Not Adherent

Another common step in PBPM is *bucketing* [28, 35]. As some cases miss out on activities and data, predictive performance becomes worse due to missing features. A simple example from the ECM domain would be the feature 'purchase order available'. A model trained with the feature being either 'available' or 'not available' could infer that its existence indicates a soon to be implemented EC, while a lack of it indicates a missing process step and a late EC. However, at the start of the EC process, this information is normally not available, and the model should not expect it. Nevertheless, even though the data set has not all dynamic features available, predicting with less features is necessary for setting early countermeasures [169]. As a result, the availability of some specific features in later process phases might result in a different outcome estimations and predictive performance. To solve this issue, PBPM suggests grouping cases into buckets by similarity.

Following truncation and bucketing, each trace σ_i is encoded as a feature vector $g_i = (g_{i1}, g_{i2} \ldots g_{ih}, label_i)$ [28]. Specifically for the dynamic features, several encoding approaches have been tested, although none dominates others. The most common encoding techniques are aggregation-based encoding and index-based encoding [35]. Aggregation-based encoding is a numerical count where every occurrence of an activity increases the value of the corresponding feature by one. The non-occurrence of an activity is encoded as zero. Compared to this simple form, index-based encoding records events as categorical features in the order of their occurrence.

Regardless of the method, static data and the true target labels are added during training. Subsequently, the data is applied as input data for training ML models. Within PBPM, various comparative studies already exist discussing which algorithm performs best. Random Forest (RF) [174], XGBoost (XGB) [175], and long-short term memory neural networks (LSTM) [176] have been shown as the most promising algorithms for process outcome prediction [35, 171].

For model evaluation, categorical PBPM applies common methods also found in regular ML research: Accuracy, F1-Score, or AUC-ROC [28]. However, in PBPM another aspect requires investigation: predictive performance over time. As dynamic features are added, predictive performance is expected to change. Following evaluation, the best model for each specific bucket is determined, which can then be used during the online phase for predicting running cases.

6.3.3 Research Objectives and Related Literature

Based on the previously discussed theoretical background, we can derive three research objectives towards an DSR artefact targeted at predicting the scheduling adherence of ECs. The first objective specifically addresses the 'earlier' goal of ECM, as the cost of averting undesired process outcomes increases with delayed intervention.

The second objective is derived from the peculiarities of PBPM. It is safe to assume that the prediction of the EC process outcome, just as with other business processes, would change as dynamic features are added. To not put further strain on the administrative capacity due to on/off false alarms, a monitoring artefact would therefore also require a sensible amount of stability in its prediction score.

The third and final objective is to determine the best algorithm and encoding combination for application. In case the same algorithm and encoding combination performs best at early prediction and TS, then it is an obvious decision. However, in any other case, it would necessitate discussion on which algorithm is best for the specific business case

Table 6.2 Overview of literature on existing methods for EC schedule adherence prediction

Contribution		Applied method	Objective 1: Early prediction of off-schedule cases	Objective 2: High stability of predicted score	Objective 3: Determining best algorithm-encoding combination
ECM Domain	[109]	ML for change propagation during EC process	- (single point)	-	✓
	[104]	ML for EC effort prediction	- (single point)	-	-
OR Domain	[194]	Comparative ML study on late parts delivery	✓*	-	✓
	[196]	ML based manufacturing processing time prediction and explainability	- (single point)	-	-
PBPM Domain	[167]	LSTM for predicting remaining time in process	-	-	✓
	[170]	Comparison of ML application and cost models	✓*	✓	✓
	[35]	Comparative PBPM study of LSTM, XGB and RF	-	-	✓
	[171]	Comparative PBPM study of LSTM, SVM, DNN and RF	✓*	-	✓

ML = Machine Learning, LSTM = long-short-term-memory neural net, SVM= Support Vector Machine, XGB = XGBoost, RF = Random Forest,
* While these publications discuss the prediction of early off-schedule cases, it is not applied for the EC domain and its specifics.

Thus, we can state the three objectives as (Fig. 6.1):

- **Objective 1:** Early prediction of off-schedule ECs.
- **Objective 2:** High stability of the predicted score.
- **Objective 3:** Determining the best algorithm—encoding combination.

Within the body of knowledge, as seen in Table 6.2, none of the existing approaches in ECM, operations research, or PBPM has addressed all three objectives. Within the research domain of ECM only few supervised ML artefacts exist [44]. Whereas PAN AND STARK [109] apply ML for predicting the impact of ECs (i.e. change propagation), RIESENER ET AL. [104] focus on estimating EC implementation effort during approval. Both apply the ML method at a specific time in the process and have not discussed how their predictive performance could be influenced by the EC process. However, they have shown that ML operations in the ECM domain perform differently than expected from other similar cases. Furthermore, they suggest investigating whether the prediction could be frontloaded to earlier process steps.

ECM can also be interpreted as a sub-domain of the wider operations research domain, which has recently seen several applications of ML and explainable AI for early prediction of process outcomes. MEHDIYEV ET AL. [196] for instance conducted a case study to compare the prediction of processing times at a manufacturer. Their results indicate, that XGB performs best overall, although the difference to logistic regression and quantile regression forests (an extension of RF) is not significant. Similar results have also been shown by STEINBERG ET AL. [194], who applied ML for early prediction of parts availability at three distinct prediction points. Both contributions in the operations research discussed feature importance, although not distinguishing between static and dynamic data.

Furthermore, general PBPM research by KRATSCH ET AL. [171] and TEINEMAA ET AL. [170] investigated effects of case and process length towards predictive performance. It was shown that for long processes (n events > 10), deep learning algorithms such as LSTM networks perform better than shallow learners (i.e. RF and XGB). However, this performance improvement comes at an increased computational cost during training [167, 182] and requires extensive data. To add, only few publications discuss how the prediction score varies by process length. Whereas one study by KRATSCH ET AL. [171] identifies a steady decrease in Accuracy and F1-Score, most other research shows increasing AUC-ROC values with increasing prefix length [35, 167, 170, 197]. The increase in performance is rather noteworthy, as some event logs show a steep increase from approx. 0.5 to 0.9 AUC-ROC after a specific process length. In the closely related time series prediction research, this is solved by investigating a minimum prediction length [198] after which the prediction performance is good enough or does not change anymore. However, as shown by TEINEMAA ET AL. [170] this minimal length is not generalizable for each domain and business process. Here, no ML-approach has shown significant advantages over any of the others, with scores varying by 0.03 score-points on average.

To summarise, in ECM it is significantly easier to execute a countermeasure if issues are identified in early process steps [4]. However, early prediction bears a higher risk of unnecessary double checking or changing prediction values. Furthermore, research has yet to demonstrate, which algorithm performs best at predicting schedule adherence in the ECM domain. Thus, we apply PBPM methods on a given set of historic real-world ECs. Furthermore, by varying the encoding technique, we test the resilience of the learning algorithms to different feature engineering approaches.

6.4 Material and Methods

For setting up our experiments, we distinguish our predictive models by encoding technique, bucketing group, and applied algorithm. As evaluation criteria, we consider AUC-ROC and Temporal Stability. With PBPM being the guiding method, the following section is structured according to the workflow shown in Fig. 6.2.

6.4.1 Database

For our research, we consider a real-world event log of ECs processed in a 2-year period (01.01.2021—31.12.2022) at an automotive manufacturer. As we were only interested in those ECs which were implemented, cancelled or still running ECs were removed from the data set. This left a total of 111,367 ECs introduced to production in 8197 EC batches. In favour of data-based ECM, the administrative system saw a recent upgrade and introduced two main new features: Firstly, it integrates all activities and communication surrounding an EC, resulting in a thorough representation of the EC process. Secondly, the ECM system is designed as an end-to-end system, meaning that all phases of the EC process are documented in one database. As a result, the creation of an EC, the planned effectivity date, and its actual effectivity date at the assembly line are also automatically recorded.

By analysing the sequence of events in the traces, we were able to identify 9404 unique process paths which were taken from the creation of the EC to its implementation. Overall, including the starting trigger event, but excluding the closing event, 37 different activities can occur. The shortest traces consist of 2 events, while the longest EC took 95 events. The median sequence length was 15 events, while the average was 17.1. On average, 14 different activities were executed per instance. The resulting event-to-activity ratio is 14/17, indicating loops in the process. Analysing the distribution of case length (Fig. 6.3), it is observed that most ECs are finished after 20 events.

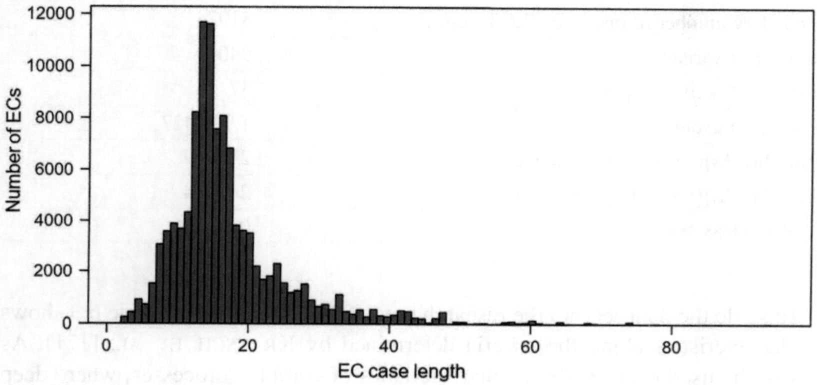

Fig. 6.3 Distribution of ECs by case length. The data indicates a left-leaning Poission distribution, with a peak at 13 and 14 events (no cases of length 1)

Regarding its variant-to-instance ratio it is difficult to state a definitive value. Considering every single EC as an instance, the ratio would be ~ 11.8, indicating a rather repetitive process. However, as there are more process paths taken than EC batches implemented, it is evident that even ECs of the same EC batch do not follow the same process. A detailed analysis reveals that of the 9404 paths, only 768 are shared across multiple EC batches, averaging on 0.74 EC batches per process path. The most common process path is shared among 140 EC batches. Furthermore, one would expect that all ECs in one EC batch are processed simultaneously and according to the same process, as this would ensure the best outcome with least resources invested [63]. As there are several contrary cases, we conclude that although instance-to-variant ratio is comparatively high on EC level, the overall EC process is highly complex. Most EC batches are processed according to an individual process, indicating that although a predefined workflow exists, it is rarely followed.

Table 6.3 Descriptive statistics of the event log

Statistic	Value
Number of instances (ECs)	111,367
Secondary number of instances (EC batches)	8197
Number of variants	9404
Number of activity types	37
Number of events	1,905,427
Min/Max/Avg. events per instance	2/95/17
Min/Max/Avg. activities per instance	2/30/14
Positive class ratio	0.83

To settle the data set into the research area of PBPM literature, Table 6.3 shows its characteristics along the criteria determined by KRATSCH ET AL. [171]. As shown, the used data set shows characteristics of complex processes, where deep learning algorithms should perform significantly better than shallow learners.

6.4.2 Prefix Extraction

In a first step, for each case the target true label was extracted. As the target label, we applied a modification of the Boolean *EC batch adherence* metric [7]. We extend the concept of EC batch adherence to individual ECs and their planned EC effectivity date. For our research, we use the definition given by BURGGRÄF ET AL. [164] as *'the last date a pre-change component is consumed in production'* as it matches the assembly perspective in a series manufacturing environment. In this scenario, it can either be a single specific date or a time frame for less critical ECs [24]. Therefore, an EC which effectivity date was in the given time frame was labelled as an 'Adherent' case, whereas ECs which were introduced outside their given time frame are labelled as 'Not Adherent'. Cases which had been ongoing, rejected, or cancelled were removed in the previous step. Furthermore, all prefixes were truncated to a maximum of 29 events, as 90% of all cases had ended by then (Fig. 6.3).

Next, basic feature engineering was conducted. This encompasses discarding rare events as well as synthesising features common to all buckets. For all cases all possible features were extracted from the ECM system and split into static features and dynamic features. Static features were initially cleaned by availability and occurrence. Boolean, categorical, and numerical features which occurred with less than 5% of cases were dismissed. Empty Boolean values were not encountered. Empty categorical features (value = null) were substituted as 'missing value'. For categorical features, all values which were only available with 5% of cases were grouped as 'other'. Empty numeric values (e.g. number of test components ordered) were substituted with 0 where appropriate. For dynamic features, the same procedure was applied. All activities which occurred with less than 5% of the instances were discarded. Unfinished activities were considered as not occurred. This resulted in a total of 30 remaining activities (excluding EC creation and EC implementation). As some features became obsolete during the investigation period, they were either dismissed or substituted if a successor feature was available.

Within PBPM literature, synthetic feature engineering has shown to improve predictive performance [35]. This is mainly done to add context and information to the cases, which is not directly available from the numerical information within the event log. For our research, we synthesised features which represent the complexity and urgency of an EC and provided them for all buckets. Complexity information was provided by calculating how many factories, products, and components are affected by the EC batch. To represent urgency and off-plan ECs, static time features (e.g. time of creation before design freeze) were added. This way, an additional 10 static features were created (see Table 6.4, 'Static Numerical Values').

Table 6.4 Static features used for predicting the outcome of the EC process

Feature	Possible Values	Level	Description
Static Categorical Values[a]			
Cost increase EC	{1,0}	Batch	indicates whether the EC batch increases the projected budget costs
Cost decrease EC	{1,0}	Batch	indicates whether the EC batch decreases the component production costs
Initial release EC	{1,0}	Batch	indicates whether the EC batch is used for an initial release of new components
Warranty EC	{1,0}	Batch	indicates whether the EC batch fixes a warranty case
Software relevant EC	{1,0}	Batch	indicates whether the EC batch affects mechatronic components
Documentation EC	{1,0}	Batch	indicates whether the EC batch fixes missing documentation
Commercially relevant EC	{1,0}	Batch	indicates whether the EC batch is commercially relevant
Quality relevant EC	{1,0}	Batch	indicates whether the EC batch is relevant for an unspecified quality issue
Delivery to external warehouse	{1,0}	EC	indicates whether the component is supplied via an external logistical centre to the factory
Follow up EC	{1,0}	Batch	indicates whether the EC batch fixes a previous EC
EC scenario (Cardinality 2)	{1,2}	EC	indicates whether the EC is a scenario I or II case
Planned purchase order set manually	{1,0}	EC	indicates whether the planned purchase order date deviates from the generically predetermined
EC during new product development	{1,0}	EC	indicates whether a new product is affected or a running series product
EC process type (Cardinality 3)	{AP, CP, GP}	Batch	indicates which process components are active for the EC batch
EC batch type (Cardinality 3)	{A, BiW, HV}	Batch	indicates which activities to execute based on assembly, body in white, and high voltage components

(continued)

Table 6.4 (continued)

Feature	Possible Values	Level	Description
Factory (Cardinality 12)	{1, 2, …,12}	EC	indicates which factory is affected by the EC batch
Manufacturer country (Cardinality 18)	{DE, CZ, …}	EC	indicates the production location of the supplier of each component, values occurring less than 5% are grouped in 'other'
Detailed EC type (Cardinality 11)	{FC, GC, …}	EC	indicates the detailed EC type, e.g. drawing change, functional change etc.
Static Numerical Values			
# previous quality ECs	{0, …, n}	EC	indicates how many EC were already implemented to fix a quality issue
# concurrent ECs per batch[e]	{0, …, n}	Batch	indicates how many ECs are grouped into the same EC batch
Project maturity	{0, …, n}	EC	numeric representation of how far the project has progressed in a products lifecycle, with max value = series production
Quality severance	{1, …, 9}	EC	numeric representation of the severity of a quality issue, with 1 being life threatening issues, 9 being no issues recorded
Quality importance	{0, 1, 2}	EC	numeric representation whether the quality issue needs special attention (2), or whether it can be solved by the production crew themselves (0)
# affected models[e]	{1, …, n}	Batch	indicates how many products are affected by the EC batch
# affected factories[e]	{1, …, n}	Batch	indicates how many factories are affected by the EC batch
Manufacturer Experience[e]	{0, …, n}	EC	indicates how many EC batches were already executed by the supplier
Creation date → commercial EC approval deadline[e]	$\{-\infty, \ldots, +\infty\}^c$	Batch	indicates how many days before the commercial approval deadline the EC batch was initiated

(continued)

Table 6.4 (continued)

Feature	Possible Values	Level	Description
Creation date → last possible Effectivity Date[e]	$\{-\infty, \ldots, +\infty\}$[c]	Batch	indicates how many days before the last preferred effectivity date the EC batch was initiated
Creation date → planned purchase order date[e]	$\{-\infty, \ldots, +\infty\}$[c]	Batch	indicates how many days before purchase order availability deadline the EC batch was initiated
Creation date → planned Effectivity Date[e]	$\{0, \ldots, +\infty\}$[c]	Batch	indicates how many days before the planned effectivity date deadline the EC batch indicates initiated
Creation date → Design freeze[e]	$\{-\infty, \ldots, +\infty\}$[c]	Batch	indicates how many days before the engineering approval deadline the EC batch was initiated
Creation date → earliest possible Effectivity date[e]	$\{-\infty, \ldots, +\infty\}$[c]	Batch	indicates how many days before the earliest preferred effectivity date the EC batch was initiated
Creation date → first delivery date deadline[e]	$\{-\infty, \ldots, +\infty\}$[c]	Batch	indicates how many days before the new parts delivery deadline the EC batch was initiated

[a] Categorical feature values that occurred in less than 5 % of cases were grouped as 'other'
[b] These features are triggered or modified by an activity, thus called payload
[c] $-\infty$ can be possible, if the EC batch is initiated after the deadline for a specific project
[d] $-\infty$ can be possible in cases where specific activities are executed before the initiation of the EC batch
[e] synthetic feature

Thus, a total of 37 static features were available from the start for all buckets and encoding approaches. Due to the EC batching strategy, some data is EC specific on a component level (e.g. supplier information), while other data is shared across ECs on batch level (e.g. affected products, factories). Table 6.4 shows a full list of all static features, including their explanation. Engineering of dynamic features, especially activity related features was postponed until encoding.

6.4.3 Bucketing

Of the various bucketing approaches, two were taken into closer consideration due to the ability to represent different points in time [35]: Bucketing by *state* [195] and bucketing by *prefix-length* [199]. From ECM literature, we can deduce that even if an EC process description exists, it is rarely followed. Moreover, some milestones are not even present with all cases. Thus, 'state bucketing' would have the drawback of first generating a valid process model, which would then provide a bucket for each process path taken. Hence, we decided to discard the state method and apply the *prefix-length bucketing method.*

As all prefixes were truncated to a maximum of 29 events, cases which were finished after fewer events had the information padded with zeros as to fill in the missing information [170]. Afterwards, 28 buckets were generated by truncating each prefix after the nth event. E.g. bucket B_{10} consists of prefixes of length 10, while bucket B_{20} has all cases truncated or padded to length 20. Each bucket is then treated as a data set, which requires encoding.

6.4.4 Encoding

RQ4 intends to investigate how early EC schedule adherence prediction is influenced by process data influence. In the context of PBPM, this process data is represented by dynamic features. To add a layer of depth, encoding dynamic features by varying methods can majorly influence the predictive performance [35]. Thus, additionally to the common aggregation and index-based approaches, we compile two other encoding methods for a total of four. As encoding methods enable enhanced synthetic features, a second round of feature engineering focusing on dynamic features takes place during encoding. For all methods, we set last-state encoding as the principal rule, i.e., only the latest information was kept, rather than keeping a record of changes. The aggregation-based methods are exemplified in Table 6.5, whereas the index-based method is showcased in Table 6.6.

A*ggregation-based* (AGG) encoding is a rather simple technique, by which every occurrence of activities within a single trace is counted and aggregated. Having 30 different activities, translating to additional 30 dynamic activity features, this resulted in the feature vector g_i having 77 features for the AGG encoded data models.

The next approach, which was devised as part of the study, we term *advanced aggregation-based encoding* (AAGG). On top of the information in the AGG method, another set of 30 dynamic features was added indicating the order of last occurrence of activities within the trace. Thus, the AAGG encoded feature vector g_i had a total of 107 features.

Adding further information, features from common synthesisation were added, such as 'activity duration' and 'activity elapsed time' (counting the time elapsed since the beginning of the process instance). These are common synthesisation suggestions as seen in [28, 35, 181]. Furthermore, we added *activity deviation* and *activity elapsed to milestone X*. These features, usually part of the index-based method, were added onto the AAGG method into a new encoding method which we term *complex aggregation-based* (CAGG) encoding, which resulted in an additional eight features per possible activity, equalling a total of 357 dynamic features for the feature vector g_i before one-hot encoding.

Table 6.5 Example of aggregation based, advanced aggregation based, and complex aggregation based encoding derived from the event log in Table 6.1

Feature[b]	Aggregation Based Encoding (AGG)	Advanced Aggregation Based Encoding (AAGG)	Complex Aggregation Based Encoding (CAGG)
Unique ID	EC 1	EC 1	EC 1
Affected Model	M1, M2	M1, M2	M1, M2
Affected Plant	P1	P1	P1
Affected Component	C1, C2, C3	C1, C2, C3	C1, C2, C3
Trigger Type	Cost red.	Cost red.	Cost red.
Cost beneficial	1	1	1
Component test	OK	OK	OK
Count Activity: Create new EC request	1	1	1
Count Activity: Calc. comm. impact	1	1	1
Count Activity: Approval	1	1	1
Count Activity: Release new design	1	1	1

(continued)

Table 6.5 (continued)

Feature[b]	Aggregation Based Encoding (AGG)	Advanced Aggregation Based Encoding (AAGG)	Complex Aggregation Based Encoding (CAGG)
Count Activity:Make new tools	2	2	2
Count Activity: Test components	2	2	2
Count Activity: Issue purchase order	1	1	1
Count Activity: Order new components	1	1	1
Count Activity: Parts delivered	1	1	1
Activity step order: Create new EC request		1	1
Activity step order: Calc. comm. impact		2	2
Activity step order: Approval		3	3
Activity step order: Release new design		4	4
Activity step order: Make new tools		7	7
Activity step order: Test components		8	8
Activity step order: Issue purchase order		9	9
Activity step order: Order new components		10	10

(continued)

Table 6.5 (continued)

Feature[b]	Aggregation Based Encoding (AGG)	Advanced Aggregation Based Encoding (AAGG)	Complex Aggregation Based Encoding (CAGG)
Activity step order: Parts delivered		11	11
Event "Release new design" date → last possible Effectivity Date[b]			–43
Event "Release new design" date → creation date[b]			10
Event "Release new design" date → EC approval deadline[b]			–1
Event "Release new design" date → first delivery date[b]			–41
Event "Release new design" date → planned PO date[b]			–33
Event "Release new design" date → earliest Effectivity Date[b]			–43
Event "Release new design" date → planned Effectivity Date[b]			–43
Activity "Release new design" completion delay[b]			–2
Implement EC	NOK	NOK	NOK

[a] Except for Unique ID and Implement EC, which are the case identifier and the process result for training / predicting
[b] for a better overview, only one example (activity: release new design) is taken for showcasing the method.

As the fourth encoding method the sophisticated *complex index-based* (IDX) encoding was set-up [28]. Differing to the previous ones, IDX encoding generates a column for n number of events and adds the activity information directly into the feature matrix. The resulting feature vector has a length of $l \times p + s$, where l is the length of the prefix, p the number of features extracted from each event, and s the number of static features available from the beginning of the case. Three features are additionally extracted to those mentioned before: *activity duration, elapsed time since last event,* and *number of events* (prefix length).

For the LSTM network, the index-based encoding is slightly modified, as instead of generating an explicit 2D feature array, the data is fed in 3D. To enable comparability, this encoding method was reproduced from the studies by Tax et al. [167], Teinemaa et al. [170], and Kratsch et al. [171], as these three had used LSTM networks with modified index-based encoding. The LSTM network was thus provided with a 3D tensor of size $i \times l \times (p + s)$, where i is the number of finished cases in the data set, l is the number of events in the prefix, and $(p + s)$ the features. Static features are replicated for each event. The full dynamic features set is shown in Table 6.7.

Following application of the encoding methods, regular data pre-processing was conducted. Categorical features were one-hot-encoded and numerical features normalised.

Table 6.6 Example of index-based encoding of the event log in Table 6.1

Feature[b]	Index Based Encoding (IDX)
Unique ID	EC 1
Affected Model	M1, M2
Affected Plant	P1
Affected Component	C1, C2, C3
Trigger Type	Cost red.
Cost beneficial	1
Component test	OK
Event 1	Create new EC request
Event 2	Calc. comm. impact
Event 3	Approval
Event 4	Release new design
Event 5	Make new tools

(continued)

Table 6.6 (continued)

Feature[b]	Index Based Encoding (IDX)
Event 5 has target date	1
Event 5 last possible effectivity date	–43
Event 5 → creation date	10
Event 5 → EC approval deadline	–1
Event 5 → first delivery date	–41
Event 5 → planned PO date	–33
Event 5 → earliest effectivity date	–43
Event 5 → planned effectivity date	–43
Event 5 activity completion delay	–2
Event 5 activity duration	1
Event 5 time elapsed since last event	1
Event 5 has target date	1
Event 6	Test components
Event 7	Make new tools
Event 8	Test components
Event 9	Issue purchase order
Event 10	Order new components
Event 11	Parts delivered
Total Number of Events	11
Implement EC	NOK

[a] Except for Unique ID and Implement EC, which are the case identifier and the process result for training / predicting
[b] for a better overview, only one example (activity: release new design) is taken for showcasing the method. In the full encoding, this additional information exists for each event number

Table 6.7 Dynamic features used for predicting the outcome of the EC process

Feature	Possible Values	Level [f]	Description	Usage by Encoding Method			
				AGG	AAGG	CAGG	IDX
Dynamic Numerical Values							
Count: Activity	{1, …, n}	Batch EC	indicates how often the activity has occurred for the given case	x	x	x	
Activity step order	{1, …, n}	Batch EC	indicates the order of activities for the given case		x	x	
Event date → last possible Effectivity Date [e]	{-∞, …, +∞} [c]	Batch EC	indicates how many days before the last preferred effectivity date the activity has occurred			x	x
Event date → creation date [e]	{-∞, …, +∞} [d]	Batch EC	indicates how many days have elapsed since the initiation of the EC batch			x	x
Event date → Design freeze [e]	{-∞, …, +∞} [c]	Batch EC	indicates how many days before the EC approval deadline the activity has occurred			x	x
Event date → first delivery date [e]	{-∞, …, +∞} [c]	EC	indicates how many days before new parts delivery deadline the activity has occurred			x	x
Event date → planned purchase order date [e]	{-∞, …, +∞} [c]	EC	indicates how many days before the purchase order availability deadline the activity has occurred			x	x
Event date → earliest possible Effectivity Date [e]	{-∞, …, +∞} [c]	EC	indicates how many days before the earliest preferred effectivity date the activity was executed			x	x
Event date → planned Effectivity Date [e]	{0, …, +∞} [c]	EC	indicates how many days are left until the target effectivity date when the activity was executed			x	x
Activity completion delay [e]	{-∞, …, +∞}	Batch EC	indicates whether the activity was delayed			x	x
Activity duration [e]	{0, …, +∞}	Batch EC	indicates how long it took to execute the activity				x
Time elapsed since last event [e]	{0, …, +∞}	Batch EC	indicates how much time has passed since the last activity				x
EC process progress [b]	{1, …, n}	Batch EC	numeric representation how much the EC process has progressed, with max value = finished	x	x	x	x

(continued)

Table 6.7 (continued)

Feature	Possible Values	Level [f]	Description	Usage by Encoding Method			
				AGG	AAGG	CAGG	IDX
Commercial approval progress [b]	{1, …, n}	Batch	numeric representation how far the commercial approval process has progressed, with max value = approved	x	x	x	x
Predicted logistical costs [b]	{0, …, +∞}	EC	indicates the predicted costs for scrapping old parts	x	x	x	x
# assembly trials conducted [b]	{0, …, +∞}	EC	indicates how often the new component was tested in an assembly trial run	x	x	x	x
# of events [e]	{0, …, +∞}	EC	indicates how many events have occurred for the given prefix and EC				x
Dynamic Categorical Values[a]							
Activity has target date (Boolean) [e]	{1,0}	Batch EC	indicates if the activity itself has a deadline				x
Activity description (Cardinality 30) [e]	{PO, AP, …}	Batch EC	description of the activity, e.g. purchase order, approval, etc.	x	x	x	x
Purchase order available (Boolean) [a]	{1,0}	EC	indicates whether a purchase order is available, payload triggered by the corresponding activity	x	x	x	x
logistically relevant EC (Boolean) [a]	{1,0}	EC	indicates whether an EC is affecting logistics, payload triggered by the corresponding activity	x	x	x	x
Confirmed delivery (Boolean) [a]	{1,0}	EC	indicates whether the material planner has confirmed delivery, payload triggered by the corresponding activity	x	x	x	x
Confirmed Supplier Readiness (Boolean) [a]	{1,0}	EC	indicates whether the supplier has confirmed delivery, payload triggered by the corresponding activity	x	x	x	x
Confirmed EC coordination status (Cardinality 6) [a]	{E, E&P, …}	Batch	indicates the level of confidence of the change coordination board, e.g. engineering confirmed, engineering & production confirmed	x	x	x	x

[a] Categorical feature values that occurred in less than 5 % of cases were grouped as 'other'
[b] These features are triggered or modified by an activity, thus called payload
[c] $-\infty$ can be possible, if the EC batch is initiated after the deadline for a specific project
[d] $-\infty$ can be possible in cases where specific activities are executed before the initiation of the EC batch
[e] synthetic feature

6.4.5 Model Training

6.4.5.1 Prediction Algorithms

Table 6.8 Overview of algorithm and encoding combination

Algorithm	Encoding Technique	Approach
Random Forest	Aggregation	RF-AGG
	Advanced Aggregation	RF-AAGG
	Complex Aggregation	RF-CAGG
	Complex Index	RF-IDX
XGBoost	Aggregation	XGB-AGG
	Advanced Aggregation	XGB-AAGG
	Complex Aggregation	XGB-CAGG
	Complex Index	XGB-IDX
LSTM Neural Network	Complex Index	LSTM

To provide an exhaustive comparison we test our encoding techniques on three algorithms which have shown to perform well in PBPM. Namely, we consider the RF, XGB and an LSTM network. All three algorithms are proven in predicting the outcome of complex processes [28, 35, 167, 171], however, none has dominated in all instances. For implementation of the RF and XGB algorithms, we decided on using off-the-shelf solutions from the SciKit Learn and XGBoost libraries for Python respectively. For implementation of the LSTM network, we applied the neural net architecture designed specifically for PBPM purposes by TAX ET AL. [167] and further adapted by TEINEMAA ET AL. [170] for classification tasks. In combination with the different encoding techniques, this resulted in nine distinct approaches (Table 6.8). Other shallow learning and deep learning algorithms have been left out, as their performance is worse, or at the best equal to those algorithms discussed above (for shallow learning see [35]), for comparison of deep learning algorithms see [171, 182].

6.4.5.2 Train—Test Split and Hyperparameter Optimisation

For training and testing, the cases were split into four groups. The initial set of ECs was fragmented using a temporal split into a train set (80% of cases) and a test set (20% of cases), as in a productive environment all available data would be consumed during training to predict newly incoming cases [170]. From the training set, 30% of cases were split off by random selection. Two thirds of these cases were used as the validation set during hyperparameter optimization and training, whereas the remaining ECs were used for calibration of the final trained models.

To ensure that no data leakage between the train, validation, calibration, and test set occurs, the ECs were split along the EC batches and the EC specific data added afterwards. As each bucket had the same split, the size of the train, validation, calibration, and test data set was the same for each bucket.

For hyperparameter optimisation, 16 random search runs were executed for each approach, as it returns similar results compared to grid search at a fraction of the invested resources [184]. This approach is especially necessary to limit the computational cost for LSTM network optimisation. Furthermore, a stopping condition was implemented to reduce overfitting. In case AUC-ROC had not improved for the last five epochs (LSTM) or iterations (RF and XGBoost), all algorithms would stop the training run during hyperparameter optimisation and final model training.

As model learning is not deterministic, each hyperparameter optimisation and each final training was re-run 20 times with the same hyperparameter settings, and the average AUC-ROC evaluated. The hyperparameters with the highest average AUC-ROC value were then used as the hyperparameters for the final test runs.

6.4.6 Evaluation

To answer the RQs, three main metrics are required: AUC-ROC, TS, and feature importance. AUC-ROC was determined as the evaluation criteria for RQ1 and RQ2, as it provides threshold independent insight in predictive performance [35]. The initial receiver-operator-characteristic curve (ROC) is plotted by calculating each of its points $P(TPR, FPR)$ as a function of the threshold t, where TPR is the True Positive Rate, and FPR the False Positive Rate. Subsequently, once the ROC is obtained, the AUC-ROC is calculated with $TPR = (1 - \beta)$ and $FPR = \alpha$ by trapezoidal integration of the receiver-operator-characteristic curve,

$$AUC - ROC = \sum_{i}\{(1 - \beta_i \times \Delta\alpha) + \tfrac{1}{2}[\Delta(1 - \beta) \times \Delta\alpha]\} \tag{6.1}$$

where

$$\Delta(1 - \beta) = (1 - \beta_i) - (1 - \beta_{i-1}) \tag{6.2}$$

$$\Delta\alpha = \alpha_i - \alpha_{i-1} \tag{6.3}$$

Furthermore, the AUC-ROC was the decision criteria for hyperparameter optimisation. A perfect classifier would have an AUC-ROC value of 1, whereas a no-skill classifier achieves 0.5. As a result of the 29 buckets, 29 individual AUC-ROC values are calculated per approach, one for each EC process step.

Building on this more general evaluation, RQ2 targets whether predictive performance is dependent on case length. Ideally, a classifier would predict perfectly from the first step onwards, independent of case length. For this, each bucket was further split into groups based on case length, and the AUC-ROC for this group evaluated and plotted versus the process length. As the minimum case length is 2, This results in 28 individual graphs showing the predictive performance from the first process step to the last. Ideally, a classifier would have a high AUC-ROC for any given prefix length regardless of case length.

RQ3 refers to a PBPM specific problem statement. As a case is processed, dynamic features are modified throughout. Consequently, the predicted outcome of a case might change from a wanted to an unwanted outcome, and back again. This would result in unnecessary double checking or overlooking negative cases. Thus, it is incentivised within PBPM that the prediction score fluctuates minimally as the case progresses. To calculate and provide a single comparative value relating to this issue, TEINEMAA ET AL. [170] introduced the *TS* metric. It represents the variation within the prediction scores between the starting and terminating events and is calculated by (Equation (6.4)):

$$TS = 1 - \frac{1}{n}\sum_{i=1}^{n}\frac{1}{T_i - 1}\sum_{t=2}^{T_i}\left|\hat{y}_t^i - \hat{y}_{t-1}^i\right| \tag{6.4}$$

where n is the number of cases, T_i the total number of events for case i, and $\hat{y}_t^i$ the prediction score of the t th event of the i th case. A perfect temporally stable classifier would in this case have a value of 1. However, there are two potential paths to a temporally stable classifier: Either the classifier is indeed suited to predict the case outcome for each step with equal confidence although the input varies, or the influence of dynamic data is insignificant and has no impact on the prediction score, and maybe not even at the process. However, the latter would also infer, that no matter the effort, the process could never result in a positive outcome.

To verify that dynamic data is indispensable and can influence the process outcome, we further investigate feature importance of the shallow learning classifiers. It is a common method and indicates how much the prediction score is influenced by each feature. However, as categorical features have been one-hot-encoded during data pre-processing, the categorical features are merged back into a single feature (e.g. the importance value for feature 'factory' is the sum of the individual importance values for factory 1, factory 2, etc.). As the feature importance method is not readily available for neural networks, their interpretability is limited and thus omitted.

To evaluate statistical significance, for each applicable evaluation metric we conducted a Friedmann rank correlation [162] followed by a post-hoc Nemenyi test [163]. Firstly, the Friedman test is applied for ranking the approaches and determining whether a statistically significant difference exists within the performances of the classifiers. For that, the Friedman statistic is calculated according to Equation 6.5:

$$\mathcal{X}_F^2 = \frac{12N}{k(k+1)}\left[\sum_{j=1}^{k} R_j^2 - \frac{k(k+1)^2}{4}\right] \tag{6.5}$$

where N is the number of test runs, k the number of ML approaches being assessed, and R_j is the sum of ranks for the jth approach. If a high $\mathcal{X}_F^2$ value is computed, a statistically significant difference in the approaches is indicated, and a Nemenyi test is necessary to determine the critical difference (CD) using Equation 6.6:

$$CD = q\sqrt{\frac{k(k+1)}{6N}} \tag{6.6}$$

To summarise the evaluation criteria, a close to perfect classifier for predicting along the EC process would have an AUC-ROC score close to 1 from the start to the end of the process for all case lengths and show little to no fluctuation in prediction scores ($TS \sim 1$). However, through feature importance evaluation it should be ensured that the predicted process outcome is indeed influenced by dynamic data. To determine statistical significance, a Friedman rank and Nemenyi test are conducted.

6.5 Results

The results relating to each metric are presented in order of the research question they are targeted at. A discussion with derived observations is presented in Sect. 6.6.1, leading to the synthesis towards implications for practical application in Sects. 6.6.2 and 6.6.3.

6.5.1 Average AUC-ROC Values by Process Length

To evaluate and compare the AUC-ROC scores (Equation (6.1)), each approach was run for 20 times for every given prefix length. The resulting values were averaged and plotted in a comparative graph, shown as Fig. 6.4, whereas the exact scores can be found in Table 6.9. Furthermore, the corresponding Friedman Rank test and Nemenyi plot can be seen in Fig. 6.5.

Table 6.9 Average values* for AUC-ROC scores of each approach split by prefix length

Prefix length	RF AGG	RF AAGG	RF CAGG	RF IDX	XGB AGG	XGB AAGG	XGB CAGG	XGB IDX	LSTM
1	0.802	0.803	0.802	0.803	**0.816**	0.815	0.813	0.815	*0.495*
2	0.803	0.802	0.797	0.791	0.810	0.814	**0.816**	0.808	*0.497*
3	0.803	0.801	0.808	0.783	0.806	0.803	**0.819**	0.804	*0.506*
4	0.803	0.801	0.806	0.782	0.805	0.807	**0.818**	0.806	*0.487*
5	0.805	0.807	0.790	0.778	0.809	0.810	**0.819**	0.816	*0.511*
6	0.802	0.800	0.767	0.782	0.818	0.810	**0.822**	0.820	*0.511*
7	0.803	0.803	0.786	0.776	0.825	0.824	0.821	**0.827**	*0.496*
8	0.802	0.800	0.784	0.769	0.818	0.815	0.815	**0.821**	*0.521*
9	0.802	0.799	0.787	0.771	0.817	0.805	0.822	**0.828**	*0.520*
10	0.798	0.794	0.759	0.766	0.811	0.817	0.809	**0.827**	*0.537*
11	0.767	0.789	0.753	0.759	0.811	0.812	0.795	**0.820**	*0.487*
12	0.796	0.788	0.759	0.749	0.810	0.807	0.805	**0.820**	*0.536*
13	0.780	0.781	0.774	0.733	0.793	0.789	0.808	**0.809**	*0.511*
14	0.753	0.766	0.761	0.729	0.778	0.793	0.784	**0.800**	*0.442*
15	0.762	0.761	0.747	0.713	0.750	0.780	0.755	**0.785**	*0.543*
16	0.745	0.743	0.709	0.708	0.744	0.748	**0.770**	0.762	*0.530*
17	0.718	0.714	0.706	0.686	0.716	0.728	**0.740**	0.731	*0.579*
18	0.737	0.728	0.733	0.703	0.741	0.722	**0.763**	0.740	*0.602*
19	0.725	0.722	0.726	0.709	0.707	0.741	**0.754**	0.728	*0.454*
20	0.703	0.735	0.660	0.709	0.751	**0.767**	0.755	0.739	*0.395*
21	0.708	0.695	0.707	0.670	0.762	**0.774**	0.711	0.733	*0.591*
22	0.718	0.743	0.683	0.725	0.767	**0.773**	0.758	0.734	*0.498*
23	0.705	0.718	0.711	0.725	**0.747**	0.719	0.743	0.718	*0.572*
24	0.730	0.708	0.702	0.674	**0.749**	0.701	0.728	0.727	*0.573*
25	0.705	0.695	0.664	0.664	0.699	**0.765**	0.703	0.711	*0.501*
26	0.708	0.707	0.631	0.671	**0.734**	0.644	0.695	0.718	*0.335*
27	0.674	0.694	0.600	0.675	**0.787**	0.656	0.683	0.711	*0.537*
28	0.662	0.658	0.598	0.681	**0.703**	0.657	0.675	0.686	*0.572*
29	0.622	0.621	0.556	0.699	**0.741**	0.707	0.622	0.646	*0.372*

*Numbers in **bold** indicate highest value for the specific case length, *italic* numbers indicate lowest value

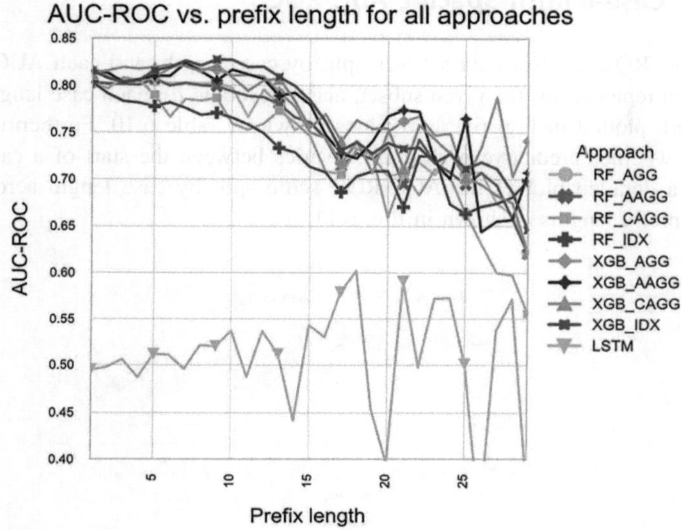

Fig. 6.4 Comparison of all 9 approaches regarding AUC-ROC vs. prefix length

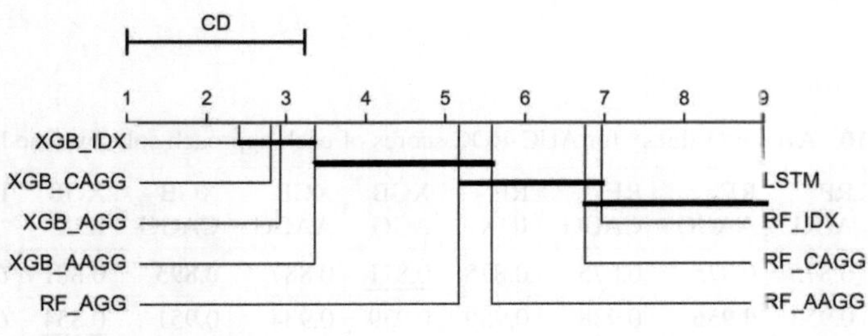

Fig. 6.5 Comparison of all approaches against each other with the Nemenyi test in terms of AUC-ROC. Groups of classifiers that are not significantly different (at $p < 0.05$) are connected

6.5.2 Case-length Specific AUC-ROC

To answer RQ2, the test data set was split by case length, and each AUC-ROC evaluation repeated for the given subset, and an average for each case length was calculated, plotted in Fig. 6.6 and values shown in Table 6.10. Furthermore, to evaluate whether predictive performance varies between the start of a case and its end, a detailed plot of the AUC-ROC score split by case length across the varying prefix lengths is shown in Fig. 6.11.

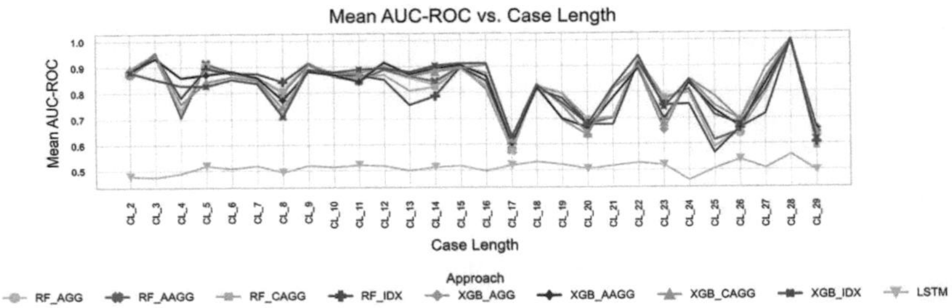

Fig. 6.6 Mean AUC-ROC scores vs. case length

Table 6.10 Average values* for AUC-ROC scores of each approach split by case length

Case length	RF AGG	RF AAGG	RF CAGG	RF IDX	XGB AGG	XGB AAGG	XGB CAGG	XGB IDX	LSTM
2	0.875	0.875	0.875	0.875	0.871	0.887	**0.895**	0.881	*0.477*
3	0.953	**0.956**	0.938	0.950	0.939	0.934	0.951	0.854	*0.473*
4	0.723	0.706	0.733	0.783	0.860	**0.860**	0.759	0.830	*0.489*
5	0.910	**0.914**	0.907	0.895	0.875	0.871	0.844	0.828	*0.519*
6	0.879	0.881	0.873	0.877	0.882	**0.884**	0.862	0.853	*0.508*
7	0.859	0.860	0.856	**0.875**	0.853	0.860	0.849	0.838	*0.519*
8	0.802	0.805	0.809	**0.843**	0.741	0.773	0.716	0.708	*0.492*
9	0.912	0.910	**0.915**	0.914	0.883	0.882	0.911	0.890	*0.519*
10	0.862	0.864	0.864	0.869	0.871	0.870	0.876	**0.880**	*0.513*
11	0.848	0.846	0.859	0.868	0.853	0.843	0.877	**0.891**	*0.521*

(continued)

Table 6.10 (continued)

Case length	RF AGG	RF AAGG	RF CAGG	RF IDX	XGB AGG	XGB AAGG	XGB CAGG	XGB IDX	LSTM
12	0.887	0.890	0.869	0.850	**0.919**	0.916	0.901	0.895	*0.517*
13	0.856	0.863	0.807	0.753	0.865	0.871	0.863	**0.879**	*0.499*
14	0.833	0.843	0.822	0.787	0.890	0.895	0.882	**0.904**	*0.512*
15	0.902	0.900	0.892	0.899	0.899	0.906	0.894	**0.912**	*0.517*
16	0.826	0.815	0.820	0.845	0.876	0.864	0.906	**0.912**	*0.496*
17	0.580	0.576	0.574	0.602	0.601	0.605	0.622	**0.630**	*0.519*
18	0.827	0.822	0.825	0.813	0.831	0.821	**0.833**	0.828	*0.533*
19	0.796	0.797	**0.797**	0.778	0.695	0.699	0.748	0.760	*0.521*
20	0.682	0.680	**0.697**	0.677	0.635	0.663	0.641	0.677	*0.504*
21	0.703	0.701	0.704	0.675	0.814	0.779	0.813	**0.821**	*0.516*
22	0.920	0.926	0.928	0.915	0.888	0.894	0.900	**0.940**	*0.527*
23	0.777	0.773	**0.781**	0.750	0.653	0.694	0.682	0.742	*0.518*
24	0.806	0.812	0.797	0.753	0.851	0.835	**0.851**	0.847	*0.459*
25	0.613	0.614	0.588	0.565	0.749	0.731	**0.784**	0.711	*0.503*
26	0.640	0.649	0.649	0.661	**0.692**	0.665	0.691	0.676	*0.539*
27	0.850	0.854	0.794	0.715	**0.889**	0.832	0.853	0.808	*0.505*
28	**1.000**	**1.000**	**1.000**	**1.000**	**1.000**	**1.000**	**1.000**	**1.000**	*0.557*
29	0.636	0.636	0.588	0.604	**0.660**	0.657	0.641	0.617	*0.494*

*Numbers in **bold** indicate highest value for the specific case length, *italic* numbers indicate lowest value, underscored values indicate the lowest shallow learner value.

6.5.3 Temporal Stability by Case Length

Given the formula in Equation (6.4) by TEINEMAA ET AL. [170], the resulting TS score was calculated for each approach and case-length. An overall average was calculated as a weighted average shown in Fig. 6.7, whereas Fig. 6.8 shows the individual TS score dependent on case length. The corresponding Friedman Rank test and Nemenyi plot whether any approach is significantly more stable than others is shown in Fig. 6.9.

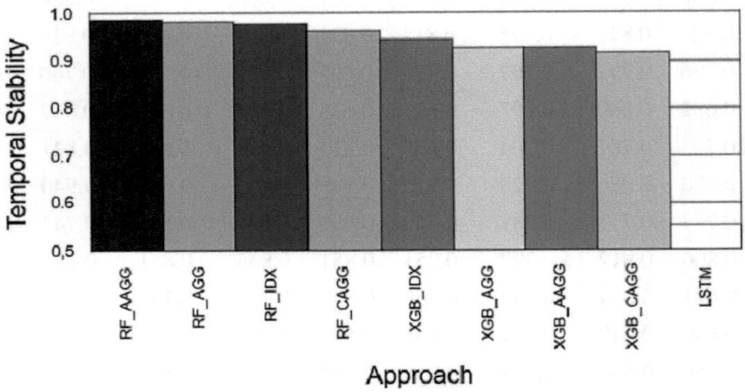

Fig. 6.7 Mean weighted temporal stability for each of the approaches, sorted from highest to lowest

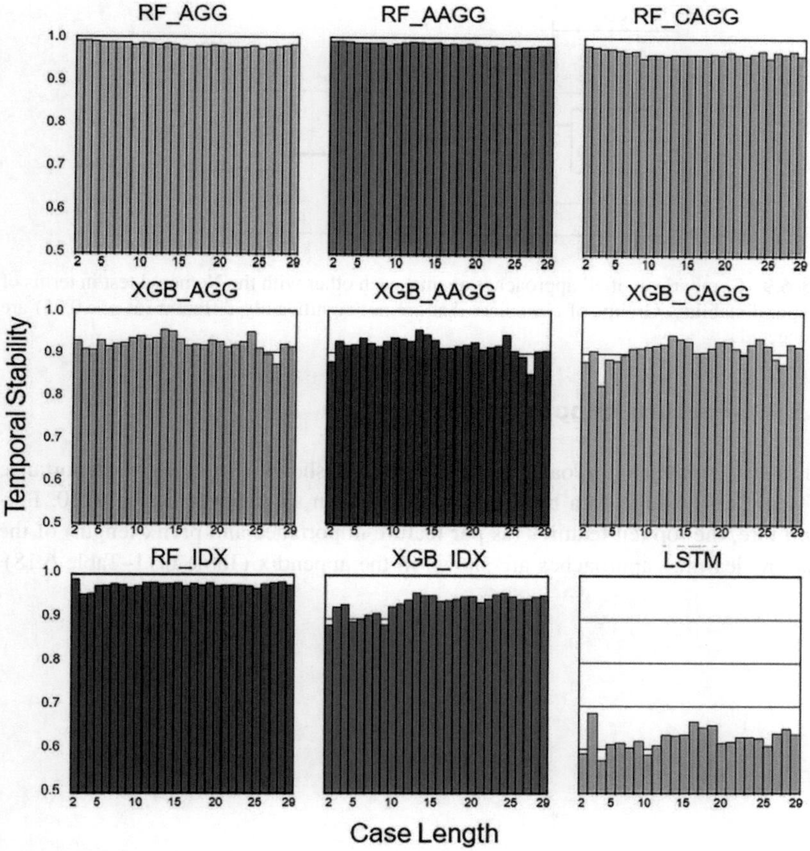

Fig. 6.8 Temporal stability for each approach split by case length

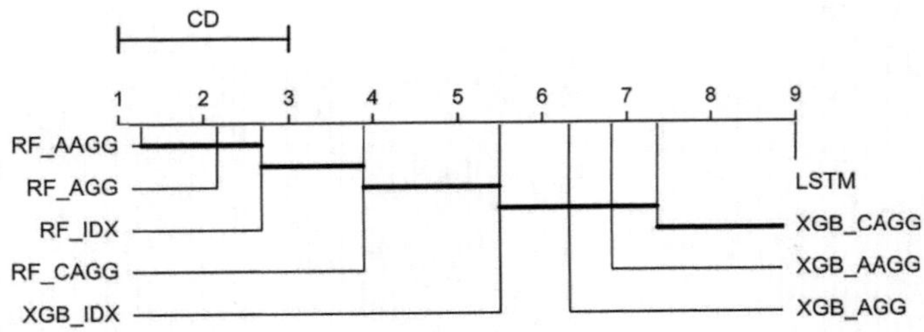

Fig. 6.9 Comparison of all approaches against each other with the Nemenyi test in terms of temporal stability. Groups of classifiers that are not significantly different (at $p < 0.05$) are connected

6.5.4 Feature Importance Evaluation

Ultimately, for each approach a graph is plotted, showing the relative importance of static to dynamic data by prefix length, shown as subplots in Fig. 6.10. Furthermore, the top ten features (as per feature importance and prefix length) of the shallow learning approaches are shown in the appendix (Table 6.11–Table 6.18)

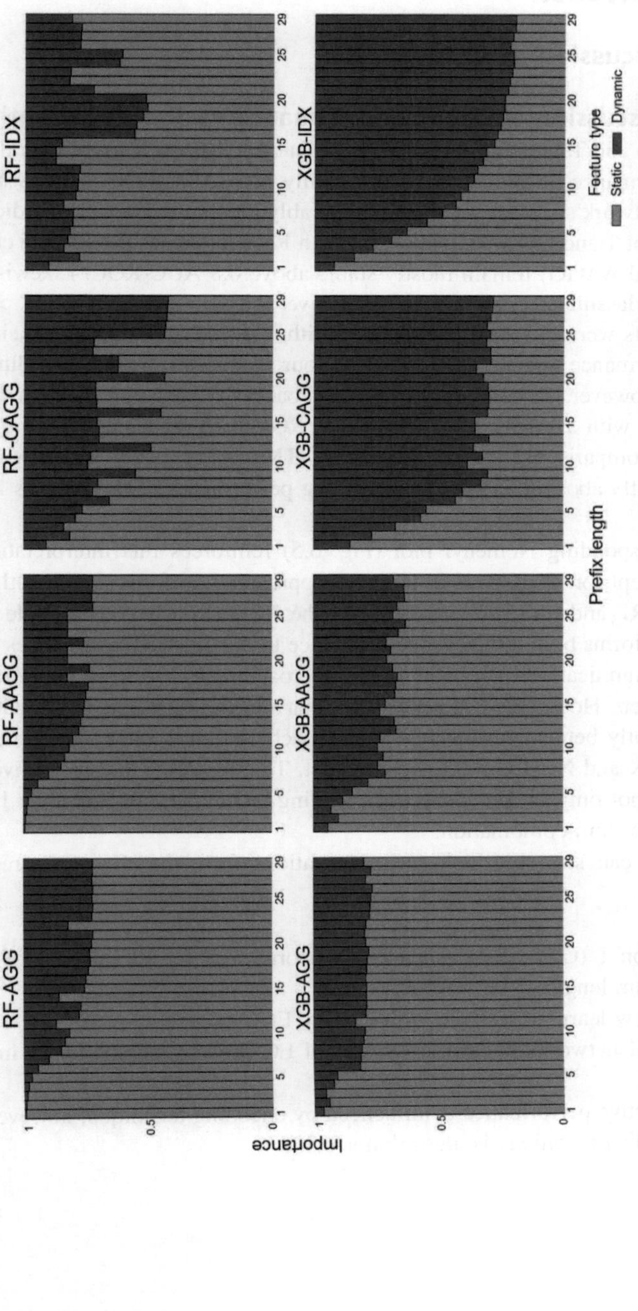

Fig. 6.10 Showcase of dynamic vs. static feature importance for each approach by prefix length Discussion and Synthesis

6.6 Evaluation

6.6.1 Discussion of Results

6.6.1.1 Discussion on Model Performance vs. Process Length

From Fig. 6.4 and Table 6.9, we can observe that for shallow learners overall predictive performance decreases from an initially high AUC-ROC score, whereas the LSTM network's increases from a comparably low base value. For predictions between Event 1 and Event 10, classifiers with RF approaches and simple encoding (AGG and AAGG) remain mostly stable above 0.8 AUC-ROC. Likewise, all XGB-based classifiers retain a good predictive performance (AUC-ROC > 0.8) until 13 events were recorded in the trace, with a period of increase in their predictive performance for prefixes of length four to ten. Following recording the 14th event, however, all shallow learning approaches decrease in their predictive performance, with a highly fluctuating AUC-ROC for prefixes of length greater than 17. In comparison, the deep learning LSTM network starts at a low AUC-ROC of slightly above 0.5, with an increasing performance as the process length increases.

The corresponding Nemenyi plot (Fig. 6.5) reinforces this interpretation of Fig. 6.4. Except for XGB-AAGG, all XGB approaches perform significantly better than the RF and LSTM approaches. Of the RF approaches, the simple AGG approach performs best, although the difference to the other RF approaches is not statistically significant. Finally, the LSTM approach underperforms, being ranked last most often. However, it is observable that the RF-IDX and RF-CAGG are not significantly better than the LSTM approach, which is rather interesting, as the XGB-IDX and XGB-CAGG perform best. This indicates that predictive performance is not only impacted by the encoding technique, but moreover by the encoding-algorithm combination.

Thus, we can state four relevant observations from the shown figures and tables:

- Observation 1 (O1): Shallow learners perform well for EC process of short and medium length.
- O2: Shallow learners struggle with lengthy EC processes.
- O3: LSTM networks underperform for all EC process lengths under investigation.
- O4: Predictive performance is influenced by encoding technique, however, the best encoding technique is algorithm specific.

From these observations, we can also answer the first research question, RQ1: For early prediction of EC effectivity date adherence, performance decreases with increasing prefix length for shallow learners, whereas it increases for LSTM networks from an initially low level.

6.6.1.2 Discussion on Model Performance vs. Case Length

Investigating the AUC-ROC performance over time for each case length (Table 6.10, Figure 6.6) reveals that all shallow learners performed similarly at predicting effectivity date adherence for several case lengths (CL 2,6,7,9,10, 11, 12, 15, and 18). It can also be observed, that although the overall predictive performance of the XGB-IDX approach seems to perform better than most other approaches (O2), it is on average the worst performing shallow learner for short cases (Table 6.10, Figure 6.6, CL 2 to 8). In fact, the previously described worse performing RF approaches outperform the XGB approaches for these shorter cases, most obviously seen with CL 5 and 8. Starting with CL 17, the performance gap between shallow learners and the LSTM network decreases. This might be attributed to two reasons: First, as seen in Fig. 6.6.3, 75% of all EC cases have finished by then, indicating a reduced amount of new information coming in. Secondly, for longer cases, more dynamic data is available, which cannot be adequately processed by shallow learners. This behaviour is consistent with other research, e.g. [170]. A special case is CL 28: Here, all shallow learners perform perfectly with an AUC-ROC of 1. Detailed investigation, however, showed that this was pure coincidence. As a result of the temporal split, there was only one batch remaining with ECs of case-length 28 in the test data, consisting of three individual ECs. Coincidentally, all shallow learners determined correctly that they would be adherent to their effectivity date.

Another observation to be drawn from detailed investigation of Fig. 6.6 is variation of predictive performance according to case progress. Especially for cases shorter than 15 events, the AUC-ROC scores vary only slightly. If plotted (see Fig. 6.11) they remain at their initial level, or slightly increase. This is rather important, as it indicates that predictive performance is good enough at the beginning of the process to roughly predict whether the EC will be on time or not. However, for cases consisting of 19 or more events, the AUC-ROC scores fluctuate heavily.

Thus, building on the observations O1-O4 we derive another set of observations from Fig. 6.6, as well as Table 6.10:

- O5: The RF classifiers perform better than XGB and LSTM classifiers at predicting effectivity date adherence of short process instances.

- O6: With increasing case-length, XGB-based approaches increase their predictive performance.
- O7: For cases of length 17 and longer (75% percentile), although average AUC-ROC values remain at good levels, the individual AUC-ROC scores for each prediction point become unstable.
- O8: With increasing prefix length, the AUC-ROC scores tend to stagnate or even increase, indicating good predictive performance at EC creation.

Regarding RQ2 we can state the following: For process instances of length 16 or shorter, the classifiers show no significant difference in predictive performance between process start and process end. Moreover, case-length specific results show that the AUC-ROC of a classifier for a specific case-length remain mostly stable for cases shorter than 16 events, regardless of prediction points.

6.6.1.3 Discussion Regarding Temporal Stability

From Fig. 6.7 we can deduce that the RF approaches are providing the most temporally stable prediction scores, i.e., they rarely change their prediction value. The corresponding Nemenyi chart (Fig. 6.9) reinforces this observation, showing that except for RF-CAGG, all RF approaches are significantly better than the XGB and LSTM approaches. Of the XGB approaches, the XGB-IDX approach performs best, although the difference is not statistically significant. Interestingly, the XGB-CAGG approach, which has performed second best at average AUC-ROC score, shows the second worst TS score. This indicates that predictive performance and TS are not necessarily correlating.

Comparing the TS values for different case lengths, we can observe that TS does not fluctuate by prefix length (Fig. 6.8). However, one interesting observation can be made for the XGB-IDX approach. Whereas all other algorithms have a high TS for short cases and a tendency for slightly lower TS for longer ones, the XGB-IDX approach tends to become more stable with longer EC processes. This behaviour can also be observed with the RF-IDX and LSTM approach (which is mostly a 3D transformed index encoding), indicating that index-based encoding enables algorithms to draw more information from dynamic data.

Likewise with the previous discussions, we draw two observations from this:

- O9: RF approaches tend to show higher TS compared to XGB and LSTM approaches, indicating different weighing of dynamic data.
- O10: A high average ROC value does not correlate to a high average TS.

For RQ3, we can state the following: The RF algorithm with AAGG encoding performs best regarding TS, however the performance difference is not significant compared to all other RF approaches.

6.6.1.4 Discussion Regarding the Relative Importance of Dynamic and Static Features

Comparing the ratio of dynamic to static data (Fig. 6.10), we can observe that relative importance of dynamic data increases for each approach as the process progresses. For the more complex approaches (IDX and CAGG), this effect becomes particularly noticeable. For the two XGB-approaches, the diminishing importance of static features almost follows a curve-function. Compared to that, the RF approaches, although showing a similar tendency, occasionally display an oscillation against the trend. Especially for the IDX and CAGG approach, the reduced importance of dynamic features could be the root cause for the comparatively weak performance.

As can be further seen by the tables Table 6.11—Table 6.18, the top 5 features are mostly the same across all eight shallow learning approaches (*Affected Model, NPD EC, Supplier Country, Project Maturity,* and *Affected Factory*). However, if we overlay this information with the relative importance of these features, we can determine that for the XGB-IDX and XGB-CAGG approaches, which perform best, the relative importance of these top 5 is comparably lower than for the other approaches, especially for prefixes longer than 10.

Thus, we draw our final two observations:

- O11: The relative importance of dynamic vs. static features increases as the process progresses, indicating that dynamic data indeed influences prediction decisions.
- O12: Although the same features are chosen among the most important, compared to XGB classifiers with complex encoding (IDX and CAGG), RF classifiers and simple XGB approaches (AGG and AAGG) overvalue the importance of static features.

As an answer to RQ4 we can thus summarise the findings as following: The relative importance of dynamic features is dependent on the algorithm and encoding combination. Whereas the XGB-IDX and XGB-CAGG approaches have the highest importance ratio of dynamic vs. static features, the corresponding RF approaches are inconsistent and shows fluctuations in the ratio.

6.6.2 Synthesis and Implications for Early Prediction Performance

Based on the twelve observations and the answers to the RQs, several implications for the early prediction of effectivity date adherence can be derived.

Observations O1, O2 and O3 suggest that shallow learners outperform the predictive performance of the LSTM network, regardless of process length. Especially for short process instances (cases finished in less than ten events), all XGB and RF classifiers outperform the LSTM network significantly. Compared to other applications of the LSTM network, the results are consistent to some degree. Teinemaa et al. [170] for instance showed, that for two out of six data sets, the LSTM network would underperform for short prefixes. Similarly, Kratsch et al. [171] research describes low AUC-ROC values of LSTM networks for one out of five data sets, attributed to imbalanced prediction classes. For our research, detailed investigation shows that underperformance can further be attributed to overfitting of the LSTM network. This might be rooted in the ECM domain and the EC batching strategy, as multiple instances share common feature values. As the LSTM network has also performed worst regarding TS (O9), we can state that the given LSTM architecture is not suited for predicting the EC effectivity dates.

From the synthesis of the answer to all four RQs, we can derive a definitive statement regarding the overarching question: Early prediction of effectivity date adherence is possible, and the XGB algorithm with index-based encoding is the most suitable candidate for operative ML models for this task. Overall, it is the best algorithm-encoding combination for predictive performance, as seen by the Nemenyi plot in Fig. 6.5. Although it is shown that for cases shorter than ten events the performance is worse than any other shallow learner's, the weighted average AUC-ROC values in Fig. 6.4 show the main reason for this assessment. As the mid-length cases (10—16 events) outnumber the shorter and longer process instances (Fig. 6.3), their results are more important. Furthermore, it is unknown how many events will be part of the case a priori. Thus, it is better to choose the classifier which performs best at a given process length, rather than for a specific case-length. Nevertheless, as the XGB-CAGG approach performs better for prefixes of length 2 to 6, it could be argued to apply this differing approach for traces of this length. As the trace length is known during prediction (count of events within the trace), this would imply a post-hoc choosing of the approach. However, this would need further investigation, as the XGB-CAGG approach has the worst TS score of all shallow learners.

An open point remains regarding cases of length 18 and longer. Here, all algorithms start to decrease and fluctuate in their predictive performance (Fig. 6.11). For these cases, it seems that the XGB-AGG and XGB-AAGG approach still perform well, with AUC-ROC values averaging around 0.72 ± 0.03 (except for Bucket B_{28}) (Table 6.9). Based on our results, no direct explanation can be found for this behaviour. From the feature importance evaluation, we observe that those two algorithms rate 'affected model' most important with a high relative importance. The unavailability of many dynamic features (from the complex encoding techniques), could further ensure that important static features are not diluted, helping the predictive performance.

6.6.3 Implications for Practical Application

From these theoretical observations and answers to the research question, we can further infer implications for practical application. Mainly, as the XGB-IDX approach performed best regarding predictive performance, whereas the RF-AAGG performed best regarding TS, choosing the best algorithm-encoding combination is dependent on organizational specifics.

Facing the industry partner with the results for the two metrics (AUC-ROC and TS), an argument can be given for both: A high AUC-ROC value implies that fewer non-adherent cases would be missed—providing a strong point for XGB-IDX. However, its lower performance for TS, would cause inefficiencies. The more temporarily stable RF-AAGG approach would come at the cost of missing more non-adherent cases, which could in turn lead to higher costs. Ultimately, the question to be answered is: Would it be better to miss detecting several unwanted outcomes or accept changing predictions and thus potential additional activities.

The industry partner decided on choosing the XGB-IDX approach as it had the best predictive performance (AUC-ROC) for cases of length 5 to 18. For cases of length 18 and more, the XGB-AGG was applied. With respect to the distribution of cases (Fig. 6.3), this meant that only a little amount of unnecessary double checking would emerge. The resulting 29 ML models based on the XGB algorithm were implemented as a web-based proof of concept tool, monitoring the predicted effectivity date for the next six months. During implementation, however, another point was raised. Predicted outcomes are not necessarily 1s or 0s (adherent and not adherent) but can also be interpreted as a likelihood to fail. This was seen as a more suitable prediction value, as associates could then decide based on their workload on which ECs to focus. Also, for cases exceeding 18 events, a separate monitoring round was set up, to ensure that long-runners and complicated cases are not lost in the overview picture. It was further decided that re-training of the ML models would take place once a month, and re-engineering of features repeated once a year or after major changes to the ECM system.

Regarding the objectives of our research (Fig. 6.1), we can conclude that all three objectives have been fulfilled:

- Regarding objective 1, we have proven that early prediction of EC schedule adherence is possible. Moreover, it was shown that it is possible to predict effectivity date adherence as soon as creating the EC with its initial static data.
- Regarding objective 2, we have shown that all shallow learners are good at providing a stable outcome. The outperforming AAGG and AGG approaches have further shown that extensive feature engineering is not always the key to stable and good predictions.
- For objective 3, we have shown that the best algorithm regarding predictive performance is not necessarily the best one regarding TS. Thus, a case sensitive decision must be taken, which was shown to be sufficient for our research case.

Finally, a question was raised: If it is possible to predict the most likely outcome, would it also be possible to prescribe the necessary activities for a positive outcome?

6.6.4 Limitations

Overall, our research has shown that prediction of adherence to effectivity dates is possible, however, there are some limitations which need to be raised.

One main limitation when transferring the results to other companies might be the availability and quality of data. At the given industry partner, the EC process was thoroughly modified and digitalised in the years leading up to the observation period. Even more so, process mining activities in the past led to a streamlined and comprehensive EC process with automatic activity recording. Most importantly, the target label to predict, 'effectivity date adherence', was automatically recorded over the past years for all ECs, establishing a good base for ML activities. As a further side result of these activities, data was readily available, and the digital documentation of the process was thorough. This is most likely not the case for all industries and companies.

Another point of limitation might be resulting from the automotive industry. As products are often iterations of a predecessor model, the number of recorded ECs is inflated. With frequent launches of new products, the number of ECs is comparably high, staff are experienced requiring less undocumented ad-hoc actions, and the software tools for administration are sophisticated. For other companies, especially engineer-to-order companies, this might not be the case and specific evaluations are necessary.

6.6.5 Future Research

For future research regarding automatic EC implementation scheduling, several research paths follow from our baseline:

Firstly, we only applied baseline algorithms and approaches. In future research, other algorithms should be tested, such as multi-model and multi-modal approaches. Multi-modal learning approaches, such as branching neural nets for instance have been shown to better manage the various data types and thus enhance predictive performance. Furthermore, we have only included semi-structured data in our research. However, in EC, a lot of information is available in the form of unstructured text, which could be leveraged by means of large language models.

Secondly, as the best algorithm-encoding combination is case-length sensitive, an investigation is necessary whether the expected case-length is predictable. This would enable a case-length sensitive prediction strategy, where based on the expected case-length, a different classifier is chosen. Towards that, regression algorithms should be tested which can indicate the remaining amount of event for any given trace.

Lastly, a move towards a true prescriptive EC scheduling system should be undertaken. Our research provides a reactive monitoring method, where the outcome is determined by the actions already taken. In a prescriptive system, the resulting tool could be used to suggest the next ideal activity and its timing so that a favourable outcome is achieved. For that, it is necessary to establish a minimum of activities so that an EC can be implemented properly. Additionally, with information on the minimum duration of these activities, their ideal sequence is calculated. Using this data, in combination with an EC schedule optimiser, a prescriptive system could compute an ideal date to implement the EC, so that costs are at a minimum and changes occur as planned.

6.7 Conclusion

Changes to products are a necessity and desired. However, during their realization, several events can lead to unwanted implementation mistakes, which lead to 'not effectivity date adherent' engineering changes. These unexpected events experienced during EC processing leads to additional complexity and reduced control. To avoid loss of profit due to late or early ECs, our research investigates whether machine learning models can be applied to predict whether an EC is effectivity date 'adherent' or 'not adherent'.

Based on observations, we can state that early prediction of effectivity date adherence is possible, regardless of process progress (i.e. process length). Among the approaches and models trained, two classifiers stood out as most promising: Combining the XGBoost learning algorithm with an index-based encoding technique resulted in the best predicting classifier measured as the AUC-ROC score. Comparably, a RF algorithm combined with an advanced aggregated encoding technique performed best regarding temporal stability. Choosing the ideal algorithm-encoding combination however is business case sensitive and should thus always be chosen based on the business goals.

Overall, we provide a baseline for predictive monitoring of EC schedule adherence. In doing so, we have shown that even outcomes of such complex processes can be predicted by applying ML. The developed and trained ML models proved applicable in a real-world environment at an industry partner.

6.8 Appendix

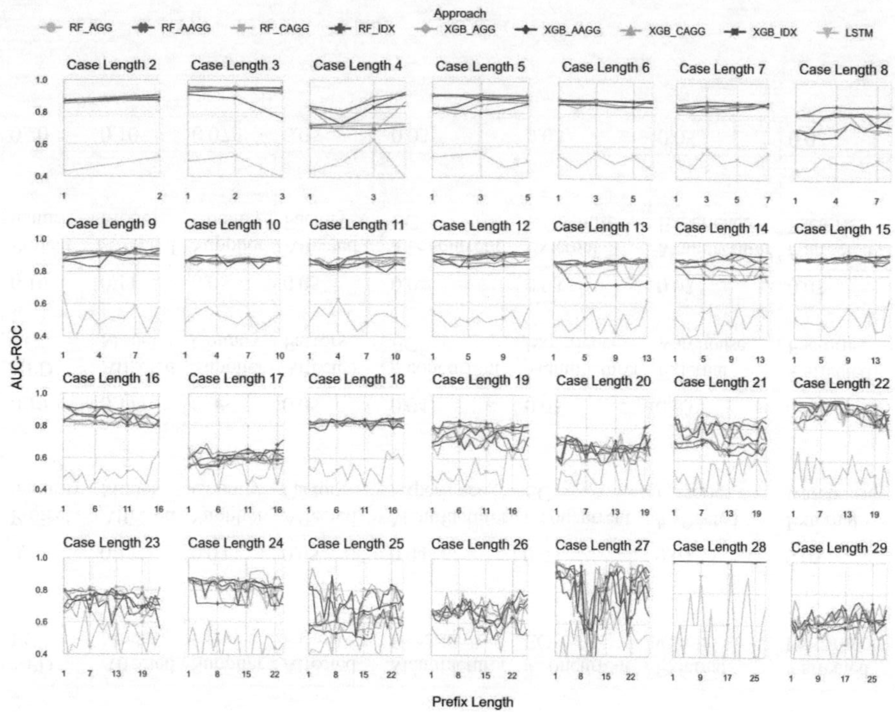

Fig. 6.11 Variation in predictive performance (measured as AUC-ROC) in dependence of case length and process length

Table 6.11 Top 10 most important features for RF-AGG by prefix length

Prefix length	Feature 1	Feature 2	Feature 3	Feature 4	Feature 5	Feature 6	Feature 7	Feature 8	Feature 9	Feature 10
1	Project maturity	NPD EC	Affected Model	Supplier Country	Affected Factory	Manufacturer Experience	# concurrent EC	External warehouse	# affected Factories	Creation to commercial approval deadline
	0.23	0.17	0.11	0.07	0.05	0.04	0.03	0.03	0.03	0.03
2	NPD EC	Project maturity	Affected Model	Supplier Country	Affected Factory	Manufacturer Experience	# concurrent EC	# affected Factories	External warehouse	Creation to commercial approval deadline
	0.21	0.19	0.11	0.06	0.05	0.04	0.03	0.03	0.02	0.02
3	Project maturity	NPD EC	Affected Model	Supplier Country	Affected Factory	# concurrent EC	Manufacturer Experience	External warehouse	# affected Factories	Creation to Design freeze
	0.21	0.19	0.11	0.06	0.05	0.04	0.03	0.03	0.03	0.02
4	NPD EC	Project maturity	Affected Model	Supplier Country	Affected Factory	# concurrent EC	External warehouse	Manufacturer Experience	# affected Factories	Creation to planned purchase order date
	0.20	0.20	0.10	0.07	0.05	0.03	0.03	0.03	0.03	0.02

(continued)

Table 6.11 (continued)

Prefix length	Feature 1	Feature 2	Feature 3	Feature 4	Feature 5	Feature 6	Feature 7	Feature 8	Feature 9	Feature 10
5	Project maturity	NPD EC	Affected Model	Supplier Country	Affected Factory	# concurrent EC	Manufacturer Experience	External warehouse	# affected Factories	Creation to commercial approval deadline
	0.20	0.20	0.11	0.06	0.05	0.03	0.03	0.03	0.03	0.02
6	Project maturity	NPD EC	Affected Model	Supplier Country	Affected Factory	Manufacturer Experience	# concurrent EC	External warehouse	# affected Factories	Creation to Design freeze
	0.21	0.19	0.10	0.06	0.05	0.03	0.03	0.03	0.03	0.02
7	Project maturity	NPD EC	Affected Model	Supplier Country	Affected Factory	# concurrent EC	External warehouse	Manufacturer Experience	# affected Factories	Creation to planned purchase order date
	0.20	0.18	0.10	0.06	0.05	0.03	0.03	0.03	0.02	0.02
8	Project maturity	NPD EC	Affected Model	Supplier Country	Affected Factory	Manufacturer Experience	External warehouse	# concurrent EC	Activity Count: CCB Buildability required	Activity: CCB Buildability confirmed
	0.18	0.17	0.10	0.06	0.05	0.03	0.03	0.03	0.03	0.02

(continued)

Table 6.11 (continued)

Prefix length	Feature 1	Feature 2	Feature 3	Feature 4	Feature 5	Feature 6	Feature 7	Feature 8	Feature 9	Feature 10
9	Project maturity	NPD EC	Affected Model	Supplier Country	Affected Factory	Activity: CCB Buildability confirmed	# concurrent EC	External warehouse	Manufacturer Experience	Activity Count: CCB Buildability required
	0.19	0.17	0.10	0.06	0.04	0.03	0.03	0.03	0.03	0.02
10	Project maturity	NPD EC	Affected Model	Supplier Country	Affected Factory	Activity: CCB Buildability confirmed	# concurrent EC	External warehouse	Activity Count: CCB Buildability required	Manufacturer Experience
	0.19	0.17	0.09	0.06	0.05	0.03	0.03	0.03	0.02	0.02
11	NPD EC	Project maturity	Affected Model	Affected Factory	Supplier Country	Activity Count: CCB Buildability required	External warehouse	Activity: CCB Buildability confirmed	# concurrent EC	Creation to first delivery
	0.27	0.20	0.10	0.07	0.05	0.03	0.03	0.03	0.02	0.01
12	Project maturity	NPD EC	Affected Model	Supplier Country	Affected Factory	Activity: CCB Buildability confirmed	External warehouse	# concurrent EC	Activity Count: CCB Buildability required	Creation to first delivery
	0.18	0.16	0.09	0.06	0.04	0.04	0.03	0.03	0.03	0.02

(continued)

Table 6.11 (continued)

Prefix length	Feature 1	Feature 2	Feature 3	Feature 4	Feature 5	Feature 6	Feature 7	Feature 8	Feature 9	Feature 10
13	Project maturity	NPD EC	Affected Model	Supplier Country	Affected Factory	Activity: CCB Buildability confirmed	# concurrent EC	Activity Count: CCB Buildability required	External warehouse	Creation to earliest effectivity date
	0.18	0.15	0.09	0.06	0.05	0.04	0.03	0.03	0.03	0.02
14	NPD EC	Project maturity	Affected Model	Affected Factory	Supplier Country	External warehouse	Activity Count: CCB Buildability required	Activity: CCB Buildability confirmed	# concurrent EC	Creation to earliest effectivity date
	0.27	0.19	0.10	0.07	0.04	0.04	0.03	0.03	0.02	0.02
15	Project maturity	NPD EC	Affected Model	Supplier Country	Affected Factory	Activity: CCB Buildability confirmed	Activity Count: CCB Buildability required	External warehouse	# concurrent EC	Creation to first delivery
	0.18	0.15	0.09	0.06	0.04	0.03	0.03	0.03	0.02	0.02
16	Project maturity	NPD EC	Affected Model	Supplier Country	Affected Factory	Activity: CCB Buildability confirmed	Activity Count: CCB Buildability required	# concurrent EC	External warehouse	Manufacturer Experience
	0.16	0.16	0.09	0.05	0.05	0.03	0.03	0.03	0.02	0.02

(continued)

Table 6.11 (continued)

Prefix length	Feature 1	Feature 2	Feature 3	Feature 4	Feature 5	Feature 6	Feature 7	Feature 8	Feature 9	Feature 10
17	Project maturity	NPD EC	Affected Model	Supplier Country	Affected Factory	Activity: CCB Buildability confirmed	Activity Count: CCB Buildability required	# concurrent EC	External warehouse	Manufacturer Experience
	0.17	0.15	0.09	0.05	0.05	0.03	0.03	0.03	0.03	0.02
18	Project maturity	NPD EC	Affected Model	Supplier Country	Affected Factory	Activity: CCB Buildability confirmed	Activity Count: CCB Buildability required	External warehouse	# concurrent EC	Activity Count: Reminder CCB Buildability
	0.16	0.15	0.09	0.05	0.04	0.03	0.03	0.03	0.03	0.02
19	Project maturity	NPD EC	Affected Model	Supplier Country	Affected Factory	Activity: CCB Buildability confirmed	External warehouse	Activity Count: CCB Buildability required	# concurrent EC	Activity: CCB Buildability required
	0.16	0.15	0.09	0.05	0.05	0.03	0.03	0.02	0.02	0.02
20	Project maturity	NPD EC	Affected Model	Supplier Country	Affected Factory	Activity: CCB Buildability confirmed	Activity Count: CCB Buildability required	External warehouse	Activity: CCB Buildability required	# concurrent EC
	0.16	0.15	0.09	0.05	0.05	0.03	0.03	0.02	0.02	0.02

(continued)

Table 6.11 (continued)

Prefix length	Feature 1	Feature 2	Feature 3	Feature 4	Feature 5	Feature 6	Feature 7	Feature 8	Feature 9	Feature 10
21	Project maturity	NPD EC	Affected Model	Supplier Country	Affected Factory	Activity: CCB Buildability confirmed	Activity: CCB Buildability required	# concurrent EC	External warehouse	Activity Count: CCB Buildability required
	0.16	0.14	0.09	0.05	0.05	0.03	0.02	0.02	0.02	0.02
22	NPD EC	Project maturity	Affected Model	Affected Factory	External warehouse	Supplier Country	Activity Count: CCB Buildability required	Activity: CCB Buildability confirmed	Activity Count: Reminder CCB Buildability	# concurrent EC
	0.25	0.19	0.10	0.07	0.04	0.04	0.03	0.03	0.02	0.02
23	Project maturity	NPD EC	Affected Model	Supplier Country	Affected Factory	Activity: CCB Buildability confirmed	External warehouse	Activity Count: Reminder CCB Buildability	Activity Count: CCB Buildability required	# concurrent EC
	0.17	0.14	0.09	0.05	0.05	0.03	0.03	0.02	0.02	0.02
24	Project maturity	NPD EC	Affected Model	Supplier Country	Affected Factory	Activity: CCB Buildability confirmed	Activity: CCB Buildability required	External warehouse	Activity Count: Reminder CCB Buildability	# concurrent EC
	0.16	0.14	0.09	0.05	0.05	0.03	0.02	0.02	0.02	0.02

(continued)

Table 6.11 (continued)

Prefix length	Feature 1	Feature 2	Feature 3	Feature 4	Feature 5	Feature 6	Feature 7	Feature 8	Feature 9	Feature 10
25	Project maturity	NPD EC	Affected Model	Supplier Country	Affected Factory	Activity: CCB Buildability confirmed	External warehouse	Activity Count: CCB Buildability required	Activity Count: Reminder CCB Buildability	# concurrent EC
	0.16	0.14	0.09	0.05	0.05	0.03	0.03	0.03	0.02	0.02
26	Project maturity	NPD EC	Affected Model	Supplier Country	Affected Factory	Activity: CCB Buildability confirmed	External warehouse	Activity Count: Reminder CCB Buildability	Activity: CCB Buildability required	# concurrent EC
	0.17	0.14	0.09	0.05	0.05	0.03	0.03	0.03	0.03	0.02
27	Project maturity	NPD EC	Affected Model	Supplier Country	Affected Factory	Activity: CCB Buildability required	Activity Count: Reminder CCB Buildability	External warehouse	# concurrent EC	Activity: CCB Buildability confirmed
	0.16	0.14	0.09	0.05	0.05	0.03	0.03	0.03	0.02	0.02

(continued)

Table 6.11 (continued)

Prefix length	Feature 1	Feature 2	Feature 3	Feature 4	Feature 5	Feature 6	Feature 7	Feature 8	Feature 9	Feature 10
28	Project maturity	NPD EC	Affected Model	Supplier Country	Affected Factory	Activity: CCB Buildability required	Activity Count: Reminder CCB Buildability	External warehouse	# concurrent EC	Activity: CCB Buildability confirmed
	0.16	0.15	0.09	0.05	0.05	0.03	0.03	0.02	0.02	0.02
29	Project maturity	NPD EC	Affected Model	Supplier Country	Affected Factory	Activity: CCB Buildability required	Activity Count: Reminder CCB Buildability	External warehouse	# concurrent EC	Activity: CCB Buildability confirmed
	0.16	0.15	0.09	0.05	0.05	0.03	0.03	0.03	0.02	0.02

Table 6.12 Top 10 most important features for RF-AAGG by prefix length

Prefix length	Feature 1	Feature 2	Feature 3	Feature 4	Feature 5	Feature 6	Feature 7	Feature 8	Feature 9	Feature 10
1	Project maturity	NPD EC	Affected Model	Supplier Country	Affected Factory	Manufacturer Experience	# concurrent EC	External warehouse	# affected Factories	Creation to comm. approval deadline
	0.23	0.17	0.11	0.07	0.05	0.04	0.03	0.03	0.03	0.03
2	NPD EC	Project maturity	Affected Model	Supplier Country	Affected Factory	Manufacturer Experience	# concurrent EC	# affected Factories	External warehouse	Creation to comm. approval deadline
	0.21	0.19	0.11	0.06	0.05	0.04	0.03	0.03	0.02	0.02
3	Project maturity	NPD EC	Affected Model	Supplier Country	Affected Factory	# concurrent EC	Manufacturer Experience	External warehouse	# affected Factories	Creation to Design freeze
	0.21	0.19	0.11	0.06	0.05	0.04	0.03	0.03	0.03	0.02
4	NPD EC	Project maturity	Affected Model	Supplier Country	Affected Factory	# concurrent EC	External warehouse	Manufacturer Experience	# affected Factories	Creation to planned purch. order date
	0.20	0.20	0.10	0.07	0.05	0.03	0.03	0.03	0.03	0.02

(continued)

Table 6.12 (continued)

Prefix length	Feature 1	Feature 2	Feature 3	Feature 4	Feature 5	Feature 6	Feature 7	Feature 8	Feature 9	Feature 10
5	Project maturity	NPD EC	Affected Model	Supplier Country	Affected Factory	# concurrent EC	Manufacturer Experience	External warehouse	# affected Factories	Creation to comm. approval deadline
	0.20	0.20	0.11	0.06	0.05	0.03	0.03	0.03	0.03	0.02
6	Project maturity	NPD EC	Affected Model	Supplier Country	Affected Factory	Manufacturer Experience	# concurrent EC	External warehouse	# affected Factories	Creation to Design freeze
	0.21	0.19	0.10	0.06	0.05	0.03	0.03	0.03	0.03	0.02
7	Project maturity	NPD EC	Affected Model	Supplier Country	Affected Factory	# concurrent EC	External warehouse	Manufacturer Experience	# affected Factories	Creation to planned purch. order date
	0.20	0.18	0.10	0.06	0.05	0.03	0.03	0.03	0.02	0.02
8	Project maturity	NPD EC	Affected Model	Supplier Country	Affected Factory	Manufacturer Experience	External warehouse	# concurrent EC	Activity Count: CCB Buildability required	Activity: CCB Buildability confirmed
	0.18	0.17	0.10	0.06	0.05	0.03	0.03	0.03	0.03	0.02

(continued)

Table 6.12 (continued)

Prefix length	Feature 1	Feature 2	Feature 3	Feature 4	Feature 5	Feature 6	Feature 7	Feature 8	Feature 9	Feature 10
9	Project maturity	NPD EC	Affected Model	Supplier Country	Affected Factory	Activity: CCB Buildability confirmed	# concurrent EC	External warehouse	Manufacturer Experience	Activity Count: CCB Buildability required
	0.19	0.17	0.10	0.06	0.04	0.03	0.03	0.03	0.03	0.02
10	Project maturity	NPD EC	Affected Model	Supplier Country	Affected Factory	Activity: CCB Buildability confirmed	# concurrent EC	External warehouse	Activity Count: CCB Buildability required	Manufacturer Experience
	0.19	0.17	0.09	0.06	0.05	0.03	0.03	0.03	0.02	0.02
11	Project maturity	NPD EC	Affected Model	Supplier Country	Affected Factory	Activity: CCB Buildability confirmed	External warehouse	Activity Count: CCB Buildability required	# concurrent EC	Manufacturer Experience
	0.19	0.16	0.10	0.06	0.05	0.04	0.03	0.03	0.03	0.02
12	Project maturity	NPD EC	Affected Model	Supplier Country	Affected Factory	Activity: CCB Buildability confirmed	External warehouse	# concurrent EC	Activity Count: CCB Buildability required	Creation to first delivery
	0.18	0.16	0.09	0.06	0.04	0.04	0.03	0.03	0.03	0.02

(continued)

Table 6.12 (continued)

Prefix length	Feature 1	Feature 2	Feature 3	Feature 4	Feature 5	Feature 6	Feature 7	Feature 8	Feature 9	Feature 10
13	Project maturity	NPD EC	Affected Model	Supplier Country	Affected Factory	Activity: CCB Buildability confirmed	# concurrent EC	Activity Count: CCB Buildability required	External warehouse	Creation to earliest effectivity date
	0.18	0.15	0.09	0.06	0.05	0.04	0.03	0.03	0.03	0.02
14	Project maturity	NPD EC	Affected Model	Supplier Country	Affected Factory	Activity: CCB Buildability confirmed	# concurrent EC	External warehouse	Activity Count: CCB Buildability required	Manufacturer Experience
	0.17	0.15	0.09	0.06	0.05	0.04	0.03	0.03	0.03	0.02
15	Project maturity	NPD EC	Affected Model	Supplier Country	Affected Factory	Activity: CCB Buildability confirmed	Activity Count: CCB Buildability required	External warehouse	# concurrent EC	Creation to first delivery
	0.18	0.15	0.09	0.06	0.04	0.03	0.03	0.03	0.02	0.02
16	Project maturity	NPD EC	Affected Model	Supplier Country	Affected Factory	Activity: CCB Buildability confirmed	Activity Count: CCB Buildability required	# concurrent EC	External warehouse	Manufacturer Experience
	0.16	0.16	0.09	0.05	0.05	0.03	0.03	0.03	0.02	0.02

(continued)

Table 6.12 (continued)

Prefix length	Feature 1	Feature 2	Feature 3	Feature 4	Feature 5	Feature 6	Feature 7	Feature 8	Feature 9	Feature 10
17	Project maturity	NPD EC	Affected Model	Supplier Country	Affected Factory	Activity: CCB Buildability confirmed	Activity Count: CCB Buildability required	# concurrent EC	External warehouse	Manufacturer Experience
	0.17	0.15	0.09	0.05	0.05	0.03	0.03	0.03	0.03	0.02
18	NPD EC	Project maturity	Affected Model	Affected Factory	External warehouse	Supplier Country	Activity Count: CCB Buildability required	Activity: CCB Buildability confirmed	# concurrent EC	Creation to earliest effectivity date
	0.26	0.19	0.10	0.07	0.04	0.03	0.03	0.03	0.02	0.02
19	Project maturity	NPD EC	Affected Model	Supplier Country	Affected Factory	Activity: CCB Buildability confirmed	External warehouse	Activity Count: CCB Buildability required	# concurrent EC	Activity: CCB Buildability required
	0.16	0.15	0.09	0.05	0.05	0.03	0.03	0.02	0.02	0.02
20	NPD EC	Project maturity	Affected Model	Affected Factory	External warehouse	Supplier Country	Activity Count: CCB Buildability required	Activity: CCB Buildability confirmed	# concurrent EC	Creation to earliest effectivity date
	0.25	0.19	0.10	0.07	0.04	0.04	0.03	0.03	0.02	0.02

(continued)

Table 6.12 (continued)

Prefix length	Feature 1	Feature 2	Feature 3	Feature 4	Feature 5	Feature 6	Feature 7	Feature 8	Feature 9	Feature 10
21	NPD EC	Project maturity	Affected Model	Affected Factory	Activity Count: CCB Buildability required	External warehouse	Supplier Country	Activity: CCB Buildability confirmed	Activity Count: Reminder CCB Buildability	Creation to earliest effectivity date
	0.25	0.19	0.10	0.07	0.03	0.03	0.03	0.02	0.02	0.02
22	Project maturity	NPD EC	Affected Model	Affected Factory	Supplier Country	Activity: CCB Buildability confirmed	External warehouse	Activity Count: CCB Buildability required	# concurrent EC	Activity Count: Reminder CCB Buildability
	0.19	0.17	0.09	0.05	0.05	0.04	0.03	0.03	0.02	0.02
23	NPD EC	Project maturity	Affected Model	Affected Factory	Supplier Country	External warehouse	Activity Count: CCB Buildability required	Activity: CCB Buildability confirmed	# concurrent EC	Activity Count: Reminder CCB Buildability
	0.25	0.19	0.10	0.07	0.04	0.04	0.03	0.02	0.02	0.02
24	NPD EC	Project maturity	Affected Model	Affected Factory	External warehouse	Supplier Country	Activity Count: CCB Buildability required	Activity: CCB Buildability confirmed	# concurrent EC	Activity Count: Reminder CCB Buildability
	0.24	0.20	0.10	0.07	0.04	0.04	0.03	0.02	0.02	0.02

(continued)

Table 6.12 (continued)

Prefix length	Feature 1	Feature 2	Feature 3	Feature 4	Feature 5	Feature 6	Feature 7	Feature 8	Feature 9	Feature 10
25	NPD EC	Project maturity	Affected Model	Affected Factory	Activity Count: CCB Buildability required	Supplier Country	External warehouse	Activity: CCB Buildability confirmed	# concurrent EC	Activity Count: Reminder CCB Buildability
	0.24	0.19	0.10	0.07	0.04	0.04	0.03	0.02	0.02	0.02
26	Project maturity	NPD EC	Affected Model	Supplier Country	Affected Factory	Activity: CCB Buildability confirmed	External warehouse	Activity Count: Reminder CCB Buildability	Activity: CCB Buildability required	# concurrent EC
	0.17	0.14	0.09	0.05	0.05	0.03	0.03	0.03	0.03	0.02
27	Project maturity	NPD EC	Affected Model	Supplier Country	Affected Factory	Activity: CCB Buildability required	Activity Count: Reminder CCB Buildability	External warehouse	# concurrent EC	Activity: CCB Buildability confirmed
	0.16	0.14	0.09	0.05	0.05	0.03	0.03	0.03	0.02	0.02

(continued)

Table 6.12 (continued)

Prefix length	Feature 1	Feature 2	Feature 3	Feature 4	Feature 5	Feature 6	Feature 7	Feature 8	Feature 9	Feature 10
28	Project maturity	NPD EC	Affected Model	Supplier Country	Affected Factory	Activity: CCB Buildability required	Activity Count: Reminder CCB Buildability	External warehouse	# concurrent EC	Activity: CCB Buildability confirmed
	0.16	0.15	0.09	0.05	0.05	0.03	0.03	0.02	0.02	0.02
29	Project maturity	NPD EC	Affected Model	Supplier Country	Affected Factory	Activity: CCB Buildability required	Activity Count: Reminder CCB Buildability	External warehouse	# concurrent EC	Activity: CCB Buildability confirmed
	0.16	0.15	0.09	0.05	0.05	0.03	0.03	0.03	0.02	0.02

Table 6.13 Top 10 most important features for RF-CAGG by prefix length

Prefix length	Feature 1	Feature 2	Feature 3	Feature 4	Feature 5	Feature 6	Feature 7	Feature 8	Feature 9	Feature 10
1	Project maturity	NPD EC	Affected Model	Supplier Country	Affected Factory	Manufacturer Experience	# concurrent EC	External warehouse	# affected Factories	# affected components
	0.19	0.18	0.10	0.06	0.05	0.03	0.03	0.03	0.03	0.02
2	Project maturity	NPD EC	Affected Model	Supplier Country	Affected Factory	Manufacturer Experience	# concurrent EC	External warehouse	# affected Factories	Activity duration: EC creation
	0.19	0.18	0.10	0.06	0.04	0.03	0.03	0.03	0.03	0.02
3	Project maturity	NPD EC	Affected Model	Supplier Country	Affected Factory	# concurrent EC	# affected Factories	External warehouse	Manufacturer Experience	Creation to comm. approval deadline
	0.21	0.18	0.09	0.06	0.04	0.03	0.03	0.03	0.03	0.02
4	Project maturity	NPD EC	Affected Model	Supplier Country	Affected Factory	External warehouse	# concurrent EC	Manufacturer Experience	# affected Factories	Creation to comm. approval deadline
	0.20	0.18	0.09	0.05	0.04	0.02	0.02	0.02	0.02	0.01
5	Project maturity	NPD EC	Affected Model	Supplier Country	Affected Factory	Manufacturer Experience	External warehouse	# concurrent EC	# affected Factories	Creation to comm. approval deadline
	0.19	0.16	0.08	0.05	0.04	0.03	0.03	0.03	0.02	0.02

(continued)

Table 6.13 (continued)

Prefix length	Feature 1	Feature 2	Feature 3	Feature 4	Feature 5	Feature 6	Feature 7	Feature 8	Feature 9	Feature 10
6	Project maturity	NPD EC	Affected Model	Supplier Country	Affected Factory	Manufacturer Experience	External warehouse	# concurrent EC	Event date to planned purch. order date (Process Planning)	# affected Factories
	0.18	0.16	0.08	0.05	0.03	0.03	0.02	0.02	0.02	0.01
7	Project maturity	NPD EC	Affected Model	Supplier Country	Affected Factory	Manufacturer Experience	# concurrent EC	External warehouse	Event date to comm. approval deadline (Confirm Supply Chain)	# affected Factories
	0.18	0.16	0.09	0.05	0.04	0.03	0.02	0.02	0.02	0.01
8	Project maturity	NPD EC	Affected Model	Supplier Country	Affected Factory	Manufacturer Experience	activity->zet-T0007	External warehouse	# concurrent EC	Event date to earliest possible Effectivity Date (Supply Chain Confirmed)
	0.18	0.17	0.07	0.05	0.04	0.03	0.02	0.02	0.02	0.02

(continued)

Table 6.13 (continued)

Prefix length	Feature 1	Feature 2	Feature 3	Feature 4	Feature 5	Feature 6	Feature 7	Feature 8	Feature 9	Feature 10
9	NPD EC	Project maturity	Affected Model	Supplier Country	Affected Factory	External warehouse	Activity completion delay (CCB Buildability required)	# concurrent EC	Event date to earliest possible effectivity date (CCB Buildability)	Event date to planned purch. order date (CCB Buildability)
	0.15	0.15	0.08	0.04	0.03	0.03	0.02	0.02	0.02	0.02
10	NPD EC	Project maturity	Affected Model	Affected Factory	External warehouse	Supplier Country	Activity completion delay (CCB Buildability required)	Activity duration: EC creation	# concurrent EC	Event date to planned purch. order date (CCB Buildability)
	0.29	0.17	0.10	0.06	0.03	0.03	0.02	0.02	0.01	0.01
11	NPD EC	Project maturity	Affected Model	Affected Factory	External warehouse	Supplier Country	Activity completion delay (CCB Buildability required)	Activity duration (CCB Buildability required)	# concurrent EC	Activity duration: EC creation
	0.28	0.17	0.10	0.06	0.03	0.03	0.02	0.02	0.02	0.02

(continued)

Table 6.13 (continued)

Prefix length	Feature 1	Feature 2	Feature 3	Feature 4	Feature 5	Feature 6	Feature 7	Feature 8	Feature 9	Feature 10
12	NPD EC	Project maturity	Affected Model	Supplier Country	Affected Factory	Activity completion delay (CCB Buildability required)	Activity duration (CCB Buildability required)	Event date to planned purch. order date (CCB Buildability)	External warehouse	# concurrent EC
	0.14	0.13	0.07	0.03	0.03	0.03	0.02	0.02	0.02	0.02
13	Project maturity	NPD EC	Affected Model	Supplier Country	Affected Factory	Manufacturer Experience	Event date to comm. approval deadline (ECN complete)	# concurrent EC	External warehouse	Confirmed EC coordination status
	0.17	0.16	0.08	0.05	0.04	0.03	0.02	0.02	0.02	0.01
14	NPD EC	Project maturity	Affected Model	Supplier Country	Affected Factory	Manufacturer Experience	# concurrent EC	Event date to comm. approval deadline (ECN complete)	External warehouse	Confirmed EC coordination status
	0.17	0.17	0.08	0.05	0.04	0.03	0.02	0.02	0.02	0.02
15	Project maturity	NPD EC	Affected Model	Supplier Country	Affected Factory	Manufacturer Experience	# concurrent EC	External warehouse	Event date to purch. order deadline (ECN complete)	# affected Factories
	0.18	0.17	0.09	0.05	0.04	0.02	0.02	0.02	0.01	0.01

(continued)

Table 6.13 (continued)

Prefix length	Feature 1	Feature 2	Feature 3	Feature 4	Feature 5	Feature 6	Feature 7	Feature 8	Feature 9	Feature 10
16	Project maturity	NPD EC	Affected Model	Event date to purch. order deadline (ECN complete)	Supplier Country	Affected Factory	Event date to comm. approval deadline (ECN complete)	External warehouse	Activity duration (CCB Buildability required)	Event date to planned purch. order date (CCB Buildability)
	0.13	0.12	0.06	0.04	0.03	0.03	0.03	0.02	0.02	0.01
17	Project maturity	NPD EC	Affected Model	Supplier Country	Affected Factory	Manufacturer Experience	External warehouse	# concurrent EC	# affected Factories	Event date to purch. order deadline (ECN complete)
	0.18	0.17	0.09	0.05	0.04	0.02	0.02	0.02	0.02	0.01
18	Project maturity	NPD EC	Affected Model	Supplier Country	Affected Factory	Manufacturer Experience	External warehouse	# concurrent EC	# affected Factories	Creation to planned purch. order date
	0.18	0.16	0.09	0.05	0.04	0.03	0.02	0.02	0.02	0.01
19	Project maturity	NPD EC	Affected Model	Supplier Country	Affected Factory	Manufacturer Experience	# concurrent EC	External warehouse	# affected Factories	Event date to purch. order deadline (ECN complete)
	0.18	0.17	0.09	0.05	0.04	0.02	0.02	0.02	0.02	0.01

(continued)

Table 6.13 (continued)

Prefix length	Feature 1	Feature 2	Feature 3	Feature 4	Feature 5	Feature 6	Feature 7	Feature 8	Feature 9	Feature 10
20	Project maturity	NPD EC	Affected Model	Event date to purch. order deadline (ECN complete)	Supplier Country	Event date to comm. approval deadline (ECN complete)	Affected Factory	External warehouse	Event date to planned purch. order date (CCB Buildability)	Activity completion delay (CCB Buildability required)
	0.13	0.12	0.06	0.04	0.03	0.03	0.03	0.02	0.02	0.01
21	Project maturity	NPD EC	Affected Model	Affected Factory	External warehouse	Event date to purch. order deadline (ECN complete)	# concurrent EC	Activity duration (Confirm Order)	Supplier Country	Activity duration (Design solution)
	0.23	0.14	0.08	0.05	0.04	0.02	0.01	0.01	0.01	0.01
22	Project maturity	NPD EC	Affected Model	Affected Factory	External warehouse	Event date to purch. order deadline (ECN complete)	# concurrent EC	Activity duration (Confirm Order)	Event date to comm. approval deadline (ECN complete)	Activity duration (Design solution)
	0.23	0.14	0.08	0.05	0.04	0.02	0.01	0.01	0.01	0.01

(continued)

Table 6.13 (continued)

Prefix length	Feature 1	Feature 2	Feature 3	Feature 4	Feature 5	Feature 6	Feature 7	Feature 8	Feature 9	Feature 10
23	Project maturity	NPD EC	Affected Model	Supplier Country	Affected Factory	Manufacturer Experience	External warehouse	# affected Factories	# concurrent EC	Creation to planned purch. order date
	0.18	0.17	0.09	0.05	0.04	0.03	0.02	0.02	0.02	0.01
24	Project maturity	NPD EC	Affected Model	Supplier Country	Affected Factory	Manufacturer Experience	External warehouse	# affected Factories	# concurrent EC	Creation to planned purch. order date
	0.18	0.17	0.09	0.05	0.04	0.03	0.02	0.02	0.02	0.02
25	Project maturity	NPD EC	Affected Model	Event date to purch. order deadline (ECN complete)	Supplier Country	Event date to comm. approval deadline (ECN complete)	Affected Factory	External warehouse	Activity duration (CCB Buildability required)	Activity completion delay (CCB Buildability required)
	0.12	0.12	0.06	0.04	0.04	0.03	0.03	0.02	0.02	0.01
26	Project maturity	NPD EC	Affected Model	Event date to purch. order deadline (ECN complete)	Event date to comm. approval deadline (ECN complete)	Supplier Country	Affected Factory	External warehouse	Activity duration (CCB Buildability required)	Activity completion delay (CCB Buildability required)
	0.12	0.12	0.06	0.04	0.03	0.03	0.03	0.02	0.02	0.02

(continued)

Table 6.13 (continued)

Prefix length	Feature 1	Feature 2	Feature 3	Feature 4	Feature 5	Feature 6	Feature 7	Feature 8	Feature 9	Feature 10
27	Project maturity	NPD EC	Affected Model	Event date to purch. order deadline (ECN complete)	Affected Factory	Supplier Country	Event date to comm. approval deadline (ECN complete)	External warehouse	Activity duration (CCB Buildability required)	Event date to purch. order deadline (Reminder CCB Buildability)
	0.13	0.12	0.06	0.04	0.03	0.03	0.03	0.02	0.02	0.02
28	NPD EC	Project maturity	Affected Model	Event date to purch. order deadline (ECN complete)	Supplier Country	Event date to comm. approval deadline (ECN complete)	Affected Factory	External warehouse	Activity duration (CCB Buildability required)	Event date to purch. order deadline (Reminder CCB Buildability)
	0.13	0.12	0.06	0.04	0.03	0.03	0.03	0.02	0.02	0.02
29	Project maturity	NPD EC	Affected Model	Event date to purch. order deadline (ECN complete)	Supplier Country	Event date to comm. approval deadline (ECN complete)	Affected Factory	External warehouse	Activity duration (CCB Buildability required)	Activity completion delay (Reminder CCB Buildability)
	0.12	0.12	0.06	0.04	0.03	0.03	0.03	0.02	0.02	0.02

Table 6.14 Top 10 most important features for RF-IDX by prefix length

Prefix length	Feature 1	Feature 2	Feature 3	Feature 4	Feature 5	Feature 6	Feature 7	Feature 8	Feature 9	Feature 10
1	Project maturity	NPD EC	Affected Model	Supplier Country	Affected Factory	# concurrent EC	Manufacturer Experience	External warehouse	# affected Factories	# affected components
	0.20	0.18	0.10	0.07	0.05	0.03	0.03	0.03	0.03	0.02
2	NPD EC	Project maturity	Affected Model	Supplier Country	Affected Factory	Event date to comm. approval deadline (1st activity)	Manufacturer Experience	# concurrent EC	External warehouse	# affected Factories
	0.19	0.17	0.09	0.06	0.04	0.03	0.03	0.03	0.03	0.02
3	NPD EC	Project maturity	Affected Model	Affected Factory	Supplier Country	External warehouse	# concurrent EC	Activity completion duration (1st activity)	Event date to creation (1st activity)	Manufacturer Experience
	0.29	0.21	0.11	0.05	0.05	0.04	0.02	0.01	0.01	0.01
4	NPD EC	Project maturity	Affected Model	Affected Factory	Supplier Country	External warehouse	# concurrent EC	Activity completion duration (1st activity)	Event date to creation (1st activity)	# affected Factories
	0.29	0.20	0.10	0.06	0.05	0.03	0.02	0.01	0.01	0.01

(continued)

Table 6.14 (continued)

Prefix length	Feature 1	Feature 2	Feature 3	Feature 4	Feature 5	Feature 6	Feature 7	Feature 8	Feature 9	Feature 10
5	NPD EC	Project maturity	Affected Model	Supplier Country	Affected Factory	External warehouse	# concurrent EC	Activity completion duration (1st activity)	Event date to creation (1st activity)	Manufacturer Experience
	0.29	0.20	0.10	0.05	0.05	0.03	0.02	0.01	0.01	0.01
6	NPD EC	Project maturity	Affected Model	Affected Factory	Supplier Country	External warehouse	# concurrent EC	Activity completion duration (1st activity)	Event date to creation (1st activity)	Manufacturer Experience
	0.29	0.19	0.10	0.06	0.05	0.03	0.02	0.01	0.01	0.01
7	NPD EC	Project maturity	Affected Model	Affected Factory	Supplier Country	External warehouse	# concurrent EC	Event date to creation (6th activity)	Manufacturer Experience	# affected Factories
	0.28	0.21	0.10	0.05	0.05	0.03	0.02	0.02	0.01	0.01
8	NPD EC	Project maturity	Affected Model	Affected Factory	Supplier Country	External warehouse	Event date to creation (6th activity)	# concurrent EC	Event date to purch. order deadline (1st activity)	# affected Factories
	0.34	0.21	0.11	0.06	0.04	0.04	0.02	0.01	0.01	0.01

(continued)

Table 6.14 (continued)

Prefix length	Feature 1	Feature 2	Feature 3	Feature 4	Feature 5	Feature 6	Feature 7	Feature 8	Feature 9	Feature 10
9	NPD EC	Project maturity	Affected Model	Affected Factory	Supplier Country	External warehouse	# concurrent EC	Event date to creation (6th activity)	# affected Factories	Event date to purch. order deadline (1st activity)
	0.31	0.18	0.10	0.06	0.05	0.03	0.02	0.02	0.01	0.01
10	NPD EC	Project maturity	Affected Model	Affected Factory	Supplier Country	External warehouse	# concurrent EC	Event date to creation (6th activity)	# affected Factories	Activity completion duration (1st activity)
	0.27	0.22	0.11	0.05	0.04	0.03	0.02	0.02	0.01	0.01
11	NPD EC	Project maturity	Affected Model	Affected Factory	Supplier Country	External warehouse	# concurrent EC	Event date to creation (6th activity)	# affected Factories	# affected components
	0.28	0.21	0.10	0.06	0.05	0.03	0.02	0.02	0.01	0.01
12	NPD EC	Project maturity	Affected Model	Affected Factory	Supplier Country	External warehouse	Event date to creation (6th activity)	# concurrent EC	# affected Factories	Event date to purch. order deadline (1st activity)
	0.27	0.21	0.10	0.05	0.05	0.03	0.02	0.02	0.01	0.01

(continued)

Table 6.14 (continued)

Prefix length	Feature 1	Feature 2	Feature 3	Feature 4	Feature 5	Feature 6	Feature 7	Feature 8	Feature 9	Feature 10
13	NPD EC	Project maturity	Affected Model	Supplier Country	Affected Factory	External warehouse	Confirmed EC coordination status	Event date to comm. approval deadline (3rd activity)	Manufacturer Experience	Event date to comm. approval deadline (5th activity)
	0.37	0.27	0.08	0.07	0.04	0.02	0.02	0.01	0.01	0.01
14	NPD EC	Project maturity	Affected Model	Supplier Country	Affected Factory	External warehouse	Confirmed EC coordination status	Event date to comm. approval deadline (3rd activity)	Manufacturer Experience	Event date to comm. approval deadline (5th activity)
	0.36	0.29	0.07	0.07	0.04	0.02	0.02	0.01	0.01	0.01
15	NPD EC	Project maturity	Affected Model	Supplier Country	Affected Factory	External warehouse	Event date to comm. approval deadline (3rd activity)	Event date to purch. order deadline (1st activity)4	Manufacturer Experience	Event date to comm. approval deadline (5th activity)
	0.36	0.27	0.07	0.07	0.04	0.02	0.02	0.01	0.01	0.01

(continued)

Table 6.14 (continued)

Prefix length	Feature 1	Feature 2	Feature 3	Feature 4	Feature 5	Feature 6	Feature 7	Feature 8	Feature 9	Feature 10
16	NPD EC	Project maturity	Affected Model	Supplier Country	Affected Factory	External warehouse	# concurrent EC	Manufacturer Experience	Event date to comm. approval deadline (3rd activity)	Event date to purch. order deadline (3rd activity)
	0.17	0.16	0.07	0.04	0.03	0.02	0.01	0.01	0.01	0.01
17	Project maturity	NPD EC	Affected Model	Supplier Country	Affected Factory	External warehouse	Manufacturer Experience	# concurrent EC	Event date to comm. approval deadline (3rd activity)	Event date to comm. approval deadline (2nd activity)
	0.17	0.16	0.07	0.04	0.03	0.02	0.02	0.01	0.01	0.01
18	Project maturity	NPD EC	Affected Model	Supplier Country	Affected Factory	External warehouse	Event date to creation (1st activity)	Activity description: 17th	# concurrent EC	Manufacturer Experience
	0.16	0.14	0.07	0.04	0.03	0.02	0.02	0.02	0.01	0.01
19	Project maturity	NPD EC	Affected Model	Supplier Country	Affected Factory	External warehouse	Event date to creation (1st activity)	# concurrent EC	Activity description: 17th	Event date to comm. approval deadline (1st activity)
	0.15	0.15	0.07	0.04	0.02	0.02	0.02	0.01	0.01	0.01

(continued)

Table 6.14 (continued)

Prefix length	Feature 1	Feature 2	Feature 3	Feature 4	Feature 5	Feature 6	Feature 7	Feature 8	Feature 9	Feature 10
20	Project maturity	NPD EC	Affected Model	Supplier Country	Affected Factory	External warehouse	Activity has target date (19th activity)	# concurrent EC	Activity description: 17th	Event date to comm. approval deadline (1st activity)
	0.17	0.14	0.07	0.04	0.02	0.02	0.02	0.01	0.01	0.01
21	NPD EC	Project maturity	Affected Model	Affected Factory	Supplier Country	External warehouse	# concurrent EC	Event date to purch. order deadline (1st activity)	Event date to creation (6th activity)	# affected components
	0.25	0.20	0.10	0.05	0.04	0.03	0.01	0.01	0.01	0.01
22	NPD EC	Project maturity	Affected Model	Supplier Country	Affected Factory	External warehouse	Activity description: 17th	Activity has target date (19th activity)	Event date to creation (1st activity)7	Event date to purch. order deadline (1st activity)4
	0.36	0.26	0.07	0.06	0.04	0.03	0.02	0.02	0.01	0.01
23	NPD EC	Project maturity	Supplier Country	Affected Model	Affected Factory	External warehouse	Activity description: 17th	Activity has target date (19th activity)	Event date to creation (1st activity)7	Event date to purch. order deadline (1st activity)4
	0.34	0.29	0.06	0.06	0.04	0.02	0.02	0.02	0.01	0.01

(continued)

Table 6.14 (continued)

Prefix length	Feature 1	Feature 2	Feature 3	Feature 4	Feature 5	Feature 6	Feature 7	Feature 8	Feature 9	Feature 10
24	NPD EC	Project maturity	Affected Model	Supplier Country	Affected Factory	External warehouse	# concurrent EC	Event date to creation (1st activity)7	Event date to purch. order deadline (1st activity)	Activity description: 17th
	0.21	0.16	0.08	0.05	0.05	0.03	0.01	0.01	0.01	0.01
25	Project maturity	NPD EC	Affected Model	Supplier Country	Affected Factory	External warehouse	Activity description: 17th	# concurrent EC	Event date to purch. order deadline (1st activity)	Activity has target date (19th activity)
	0.20	0.17	0.08	0.04	0.04	0.03	0.01	0.01	0.01	0.01
26	NPD EC	Project maturity	Affected Model	Affected Factory	Supplier Country	External warehouse	Event date to purch. order deadline (1st activity)	Event date to creation (6th activity)	# concurrent EC	Confirmed EC coordination status
	0.31	0.21	0.10	0.07	0.03	0.03	0.01	0.01	0.01	0.01

(continued)

Table 6.14 (continued)

Prefix length	Feature 1	Feature 2	Feature 3	Feature 4	Feature 5	Feature 6	Feature 7	Feature 8	Feature 9	Feature 10
27	NPD EC	Project maturity	Affected Model	Affected Factory	Supplier Country	External warehouse	Event date to purch. order deadline (1st activity)	Event date to creation (6th activity)	# concurrent EC	Activity has target date (19th activity)
	0.31	0.20	0.10	0.07	0.04	0.03	0.01	0.01	0.01	0.01
28	NPD EC	Project maturity	Affected Model	Affected Factory	Supplier Country	External warehouse	Event date to purch. order deadline (1st activity)4	Event date to creation (6th activity)	# concurrent EC	Confirmed EC coordination status
	0.31	0.20	0.10	0.06	0.04	0.03	0.01	0.01	0.01	0.01
29	NPD EC	Project maturity	Affected Model	Supplier Country	Affected Factory	Activity description: 17th	External warehouse	Event date to creation (1st activity)7	Activity has target date (19th activity)	Event date to purch. order deadline (1st activity)4
	0.33	0.28	0.07	0.06	0.04	0.02	0.02	0.02	0.02	0.01

Table 6.15 Top 10 most important features for XGB-AGG by prefix length

Prefix length	Feature 1	Feature 2	Feature 3	Feature 4	Feature 5	Feature 6	Feature 7	Feature 8	Feature 9	Feature 10
1	Affected Model	Supplier Country	Factory Affected	NPD EC	EC type	Project maturity	External warehouse	EC process type	EC batch type	EC scenario
	0.29	0.12	0.12	0.11	0.06	0.04	0.03	0.02	0.01	0.01
2	Affected Model	NPD EC	Factory Affected	Supplier Country	EC type	External warehouse	Project maturity	EC process type	EC scenario	EC scenario
	0.28	0.14	0.13	0.12	0.04	0.03	0.03	0.01	0.01	0.01
3	Affected Model	Supplier Country	NPD EC	Factory Affected	Project maturity	EC type	External warehouse	EC scenario	# concurrent EC	Manufacturer Experience
	0.32	0.12	0.10	0.10	0.07	0.04	0.03	0.01	0.01	0.01
4	Affected Model	Supplier Country	NPD EC	Factory Affected	Project maturity	EC type	External warehouse	EC scenario	# affected Factories	Activity: ECN distributed
	0.32	0.11	0.09	0.09	0.05	0.04	0.03	0.02	0.01	0.01
5	Affected Model	NPD EC	Factory Affected	Supplier Country	Project maturity	EC type	External warehouse	EC scenario	Confirmed EC coordination status	# concurrent EC
	0.31	0.11	0.11	0.09	0.04	0.04	0.03	0.01	0.01	0.01

(continued)

Table 6.15 (continued)

Prefix length	Feature 1	Feature 2	Feature 3	Feature 4	Feature 5	Feature 6	Feature 7	Feature 8	Feature 9	Feature 10
6	Affected Model	NPD EC	Supplier Country	Factory Affected	EC type	Confirmed EC coordination status	Project maturity	External warehouse	Activity: Supply chain confirmed	Activity: Supply chain considered
	0.24	0.12	0.10	0.10	0.05	0.03	0.03	0.02	0.01	0.01
7	Affected Model	NPD EC	Supplier Country	Factory Affected	EC type	External warehouse	Project maturity	Confirmed EC coordination status	EC process type	Activity: Approval
	0.23	0.12	0.10	0.09	0.05	0.04	0.03	0.02	0.01	0.01
8	Affected Model	NPD EC	Factory Affected	Supplier Country	EC type	Confirmed EC coordination status	External warehouse	Project maturity	EC process type	Activity: Ready for approval
	0.24	0.15	0.10	0.10	0.04	0.02	0.02	0.02	0.01	0.01
9	Affected Model	Supplier Country	NPD EC	Factory Affected	EC type	Project maturity	Confirmed EC coordination status	External warehouse	Confirmed Supplier Readiness	EC process type
	0.25	0.12	0.11	0.09	0.04	0.02	0.02	0.02	0.01	0.01

(continued)

Table 6.15 (continued)

Prefix length	Feature 1	Feature 2	Feature 3	Feature 4	Feature 5	Feature 6	Feature 7	Feature 8	Feature 9	Feature 10
10	Affected Model	NPD EC	Supplier Country	Factory Affected	EC type	Confirmed EC coordination status	Project maturity	External warehouse	Activity: CCB Buildability confirmed	Activity: Confirm EC
	0.24	0.12	0.12	0.09	0.04	0.02	0.02	0.02	0.01	0.01
11	Affected Model	NPD EC	Factory Affected	Supplier Country	Project maturity	External warehouse	EC type	Confirmed EC coordination status	EC process type	Quality importance
	0.17	0.13	0.12	0.12	0.05	0.04	0.03	0.02	0.01	0.01
12	Affected Model	NPD EC	Supplier Country	Factory Affected	EC type	Project maturity	External warehouse	Confirmed EC coordination status	EC process type	Activity: CCB Buildability confirmed
	0.23	0.12	0.11	0.09	0.04	0.02	0.02	0.02	0.01	0.01
13	Affected Model	NPD EC	Supplier Country	Factory Affected	EC type	Project maturity	Confirmed EC coordination status	External warehouse	Activity: CCB Buildability confirmed	EC process type
	0.24	0.11	0.11	0.10	0.04	0.03	0.03	0.02	0.01	0.01

(continued)

Table 6.15 (continued)

Prefix length	Feature 1	Feature 2	Feature 3	Feature 4	Feature 5	Feature 6	Feature 7	Feature 8	Feature 9	Feature 10
14	Affected Model	Supplier Country	NPD EC	Factory Affected	EC type	Project maturity	Confirmed EC coordination status	External warehouse	Activity: CCB Buildability confirmed	Activity: CCB Buildability required
	0.24	0.11	0.11	0.09	0.04	0.03	0.03	0.02	0.01	0.01
15	Affected Model	NPD EC	Supplier Country	Factory Affected	EC type	Project maturity	Confirmed EC coordination status	External warehouse	Activity: CCB Buildability confirmed	EC process type
	0.23	0.13	0.12	0.09	0.04	0.02	0.02	0.02	0.01	0.01
16	Affected Model	NPD EC	Supplier Country	Factory Affected	EC type	Project maturity	Confirmed EC coordination status	External warehouse	EC process type	Activity: Request CCB Buildability
	0.24	0.12	0.11	0.09	0.04	0.02	0.02	0.02	0.01	0.01
17	Affected Model	NPD EC	Supplier Country	Factory Affected	EC type	Confirmed EC coordination status	Project maturity	External warehouse	Activity: Confirm EC	EC process type
	0.24	0.12	0.12	0.09	0.04	0.02	0.02	0.02	0.01	0.01

(continued)

Table 6.15 (continued)

Prefix length	Feature 1	Feature 2	Feature 3	Feature 4	Feature 5	Feature 6	Feature 7	Feature 8	Feature 9	Feature 10
18	Affected Model	NPD EC	Supplier Country	Factory Affected	EC type	Confirmed EC coordination status	Project maturity	External warehouse	Activity: Confirm EC	Activity: CCB Buildability required
	0.22	0.13	0.11	0.09	0.04	0.02	0.02	0.02	0.01	0.01
19	Affected Model	NPD EC	Supplier Country	Factory Affected	EC type	Confirmed EC coordination status	Project maturity	External warehouse	Activity: Confirm EC	Activity: CCB Buildability required
	0.23	0.13	0.11	0.09	0.03	0.02	0.02	0.02	0.01	0.01
20	Affected Model	NPD EC	Supplier Country	Factory Affected	EC type	Project maturity	Confirmed EC coordination status	External warehouse	Activity: CCB Buildability required	Activity: CCB Buildability confirmed
	0.24	0.12	0.11	0.09	0.04	0.03	0.03	0.02	0.01	0.01
21	Affected Model	NPD EC	Supplier Country	Factory Affected	EC type	Project maturity	Confirmed EC coordination status	External warehouse	Activity: CCB Buildability required	Activity: CCB Buildability confirmed
	0.25	0.11	0.11	0.09	0.04	0.03	0.02	0.02	0.01	0.01

(continued)

Table 6.15 (continued)

Prefix length	Feature 1	Feature 2	Feature 3	Feature 4	Feature 5	Feature 6	Feature 7	Feature 8	Feature 9	Feature 10
22	Affected Model	NPD EC	Supplier Country	Factory Affected	EC type	Project maturity	Confirmed EC coordination status	External warehouse	Activity: CCB Buildability required	EC scenario
	0.24	0.11	0.11	0.09	0.04	0.03	0.03	0.02	0.01	0.01
23	Affected Model	NPD EC	Supplier Country	Factory Affected	EC type	Confirmed EC coordination status	External warehouse	Project maturity	Activity: CCB Buildability required	Activity: Confirm EC
	0.22	0.13	0.11	0.09	0.04	0.03	0.02	0.02	0.01	0.01
24	Affected Model	NPD EC	Supplier Country	Factory Affected	EC type	Confirmed EC coordination status	External warehouse	Project maturity	Activity: CCB Buildability required	EC process type
	0.22	0.13	0.11	0.09	0.04	0.03	0.02	0.02	0.02	0.01
25	Affected Model	NPD EC	Supplier Country	Factory Affected	EC type	Confirmed EC coordination status	External warehouse	Project maturity	Activity: CCB Buildability required	Activity: Confirm EC
	0.22	0.12	0.11	0.09	0.04	0.02	0.02	0.02	0.02	0.01

(continued)

Table 6.15 (continued)

Prefix length	Feature 1	Feature 2	Feature 3	Feature 4	Feature 5	Feature 6	Feature 7	Feature 8	Feature 9	Feature 10
26	Affected Model	NPD EC	Supplier Country	Factory Affected	EC type	Confirmed EC coordination status	Project maturity	External warehouse	Activity: CCB Buildability required	Activity: Confirm EC
	0.22	0.12	0.11	0.09	0.03	0.03	0.02	0.02	0.02	0.01
27	Affected Model	NPD EC	Supplier Country	Factory Affected	EC type	Project maturity	Confirmed EC coordination status	External warehouse	Activity: CCB Buildability required	EC scenario
	0.24	0.12	0.11	0.09	0.04	0.03	0.02	0.02	0.02	0.01
28	Affected Model	NPD EC	Supplier Country	Factory Affected	EC type	Confirmed EC coordination status	Project maturity	External warehouse	Activity: CCB Buildability required	Activity: Confirm EC
	0.22	0.12 NPD EC	0.12	0.09	0.04	0.03	0.02	0.02	0.02	0.01
29	Affected Model	Supplier Country	Factory Affected	EC type	Project maturity	Confirmed EC coordination status		External warehouse	Activity: CCB Buildability required	EC scenario
	0.25	0.11	0.11	0.09	0.04	0.03	0.02	0.02	0.02	0.01

Table 6.16 Top 10 most important features for XGB-AAGG by prefix length

Prefix length	Feature 1	Feature 2	Feature 3	Feature 4	Feature 5	Feature 6	Feature 7	Feature 8	Feature 9	Feature 10
1	Affected Model	NPD EC	Affected Factory	Supplier Country	Detailed EC type	Project maturity	External warehouse	EC batch type	Commercially relevant EC	Purchase order available
	0.26	0.14	0.12	0.12	0.07	0.03	0.02	0.01	0.01	0.01
2	Affected Model	NPD EC	Affected Factory	Supplier Country	Detailed EC type	Project maturity	External warehouse	EC batch type	Activity step order: Check part number	Activity step order: creation
	0.26	0.14	0.12	0.11	0.05	0.03	0.02	0.01	0.01	0.01
3	Affected Model	NPD EC	Affected Factory	Supplier Country	Project maturity	Detailed EC type	External warehouse	EC scenario	# concurrent EC	Creation to Design freeze
	0.32	0.12	0.11	0.09	0.04	0.04	0.03	0.01	0.01	0.01
4	Affected Model	NPD EC	Supplier Country	Affected Factory	Detailed EC type	Project maturity	External warehouse	EC scenario	# affected Factories	# concurrent EC
	0.30	0.14	0.10	0.09	0.04	0.04	0.02	0.01	0.01	0.01
5	Affected Model	NPD EC	Supplier Country	Affected Factory	Detailed EC type	Project maturity	External warehouse	EC scenario	Confirmed EC coordination status	# concurrent EC
	0.27	0.10	0.10	0.10	0.04	0.04	0.02	0.02	0.01	0.01

(continued)

Table 6.16 (continued)

Prefix length	Feature 1	Feature 2	Feature 3	Feature 4	Feature 5	Feature 6	Feature 7	Feature 8	Feature 9	Feature 10
6	Affected Model	Supplier Country	NPD EC	Affected Factory	Detailed EC type	Project maturity	External warehouse	Confirmed EC coordination status	EC scenario	Activity Count: Supply Chain Confirmed
	0.25	0.10	0.09	0.09	0.04	0.03	0.02	0.02	0.02	0.01
7	Affected Model	NPD EC	Supplier Country	Affected Factory	Detailed EC type	External warehouse	Confirmed EC coordination status	Project maturity	EC batch type	Activity Count: Approval Required
	0.19	0.12	0.10	0.08	0.03	0.03	0.02	0.02	0.01	0.01
8	Affected Model	NPD EC	Supplier Country	Affected Factory	Project maturity	Detailed EC type	Confirmed EC coordination status	External warehouse	Activity Count: Supply Chain Confirmed	Confirmed Supplier Readiness
	0.25	0.11	0.10	0.08	0.03	0.03	0.02	0.01	0.01	0.01
9	Affected Model	NPD EC	Supplier Country	Affected Factory	Project maturity	Detailed EC type	Confirmed EC coordination status	External warehouse	Activity Count: CCB Buildability required	Activity: CCB Buildability confirmed
	0.21	0.11	0.11	0.08	0.03	0.03	0.02	0.02	0.01	0.01

(continued)

Table 6.16 (continued)

Prefix length	Feature 1	Feature 2	Feature 3	Feature 4	Feature 5	Feature 6	Feature 7	Feature 8	Feature 9	Feature 10
10	Affected Model	NPD EC	Affected Factory	Supplier Country	Project maturity	External warehouse	Detailed EC type	Activity Count: CCB Buildability Required	Confirmed EC coordination status	Activity step order: Check EC batch
	0.16	0.12	0.10	0.10	0.05	0.03	0.02	0.02	0.02	0.01
11	Affected Model	NPD EC	Supplier Country	Affected Factory	Project maturity	External warehouse	Detailed EC type	Confirmed EC coordination status	Activity Count: Confirm Supply Chain	Activity Count: CCB Buildability Required
	0.16	0.12	0.11	0.10	0.05	0.03	0.02	0.01	0.01	0.01
12	Affected Model	NPD EC	Supplier Country	Affected Factory	Detailed EC type	Activity: Confirm EC	External warehouse	Confirmed EC coordination status	Project maturity	Activity Count: CCB Buildability required
	0.20	0.10	0.08	0.08	0.04	0.02	0.02	0.02	0.02	0.01
13	Affected Model	Supplier Country	NPD EC	Affected Factory	Detailed EC type	Project maturity	External warehouse	Confirmed EC coordination status	Activity step order: Design solution	Activity step order: CCB Buildability Confirmed
	0.22	0.09	0.08	0.08	0.03	0.03	0.02	0.02	0.01	0.01

(continued)

Table 6.16 (continued)

Prefix length	Feature 1	Feature 2	Feature 3	Feature 4	Feature 5	Feature 6	Feature 7	Feature 8	Feature 9	Feature 10
14	Affected Model	NPD EC	Supplier Country	Affected Factory	Project maturity	Detailed EC type	External warehouse	Confirmed EC coordination status	Activity Count: CCB Buildability Required	EC batch type
	0.16	0.11	0.10	0.10	0.05	0.03	0.03	0.02	0.01	0.01
15	Affected Model	Supplier Country	NPD EC	Affected Factory	Project maturity	External warehouse	Detailed EC type	Confirmed EC coordination status	Activity Count: CCB Buildability Required	Activity Count: Confirm Supply Chain
	0.15	0.11	0.11	0.09	0.05	0.03	0.03	0.02	0.02	0.01
16	Affected Model	NPD EC	Supplier Country	Affected Factory	Detailed EC type	Confirmed EC coordination status	External warehouse	Project maturity	Activity: Confirm EC	Activity Count: CCB Buildability Required
	0.20	0.11	0.09	0.08	0.03	0.02	0.02	0.02	0.01	0.01
17	Affected Model	NPD EC	Supplier Country	Affected Factory	Detailed EC type	Confirmed EC coordination status	External warehouse	Project maturity	Activity: Confirm EC	Activity Count: CCB Buildability required
	0.20	0.11	0.09	0.08	0.03	0.02	0.02	0.02	0.02	0.01

(continued)

Table 6.16 (continued)

Prefix length	Feature 1	Feature 2	Feature 3	Feature 4	Feature 5	Feature 6	Feature 7	Feature 8	Feature 9	Feature 10
18	Affected Model	NPD EC	Supplier Country	Affected Factory	Detailed EC type	Confirmed EC coordination status	External warehouse	Project maturity	Activity: Confirm EC	Activity Count: CCB Buildability required
	0.20	0.11	0.09	0.07	0.04	0.02	0.02	0.02	0.02	0.01
19	Affected Model	Supplier Country	Affected Factory	NPD EC	Project maturity	Detailed EC type	Confirmed EC coordination status	External warehouse	Activity Count: Reminder CCB Buildability	Activity step order: CCB Buildability Confirmed
	0.23	0.09	0.08	0.06	0.04	0.03	0.03	0.02	0.01	0.01
20	Affected Model	NPD EC	Supplier Country	Affected Factory	Project maturity	External warehouse	Detailed EC type	Activity Count: CCB Buildability Required	Confirmed EC coordination status	event-nr-T0055
	0.16	0.12	0.10	0.09	0.05	0.03	0.03	0.02	0.02	0.01
21	Affected Model	Supplier Country	Affected Factory	NPD EC	Project maturity	Detailed EC type	External warehouse	Confirmed EC coordination status	Activity Count: Reminder CCB Buildability	Activity step order: CCB Buildability Confirmed
	0.23	0.09	0.08	0.06	0.04	0.03	0.02	0.02	0.01	0.01

(continued)

Table 6.16 (continued)

Prefix length	Feature 1	Feature 2	Feature 3	Feature 4	Feature 5	Feature 6	Feature 7	Feature 8	Feature 9	Feature 10
22	Affected Model	NPD EC	Supplier Country	Affected Factory	Project maturity	External warehouse	Detailed EC type	Confirmed EC coordination status	Activity Count: CCB Buildability Required	EC batch type
	0.16	0.11	0.10	0.09	0.05	0.03	0.03	0.02	0.02	0.01
23	Affected Model	NPD EC	Supplier Country	Affected Factory	Detailed EC type	Confirmed EC coordination status	External warehouse	Project maturity	Activity: Confirm EC	Activity Count: CCB Buildability required
	0.20	0.11	0.09	0.08	0.03	0.03	0.02	0.02	0.02	0.01
24	Affected Model	NPD EC	Supplier Country	Affected Factory	Activity: CCB Buildability required	Detailed EC type	Confirmed EC coordination status	Activity: Confirm EC	External warehouse	Project maturity
	0.19	0.11	0.09	0.07	0.04	0.03	0.03	0.02	0.02	0.02
25	Affected Model	Supplier Country	Affected Factory	NPD EC	Project maturity	Detailed EC type	Confirmed EC coordination status	External warehouse	Activity Count: Reminder CCB Buildability	Activity step order: CCB Buildability Confirmed
	0.23	0.09	0.08	0.06	0.03	0.03	0.02	0.02	0.01	0.01

(continued)

Table 6.16 (continued)

Prefix length	Feature 1	Feature 2	Feature 3	Feature 4	Feature 5	Feature 6	Feature 7	Feature 8	Feature 9	Feature 10
26	Affected Model	NPD EC	Supplier Country	Affected Factory	Detailed EC type	Confirmed EC coordination status	External warehouse	Project maturity	Activity: Confirm EC	Activity Count: CCB Buildability required
	0.19	0.11	0.09	0.08	0.04	0.03	0.02	0.02	0.01	0.01
27	Affected Model	NPD EC	Supplier Country	Affected Factory	Activity: CCB Buildability required	Detailed EC type	Confirmed EC coordination status	External warehouse	Project maturity	Activity: Confirm EC
	0.18	0.10	0.08	0.08	0.04	0.04	0.02	0.02	0.02	0.02
28	Affected Model	NPD EC	Supplier Country	Affected Factory	Activity: CCB Buildability required	Detailed EC type	Confirmed EC coordination status	External warehouse	Project maturity	Activity: Confirm EC
	0.18	0.10	0.09	0.07	0.04	0.03	0.02	0.02	0.02	0.02
29	Affected Model	Supplier Country	Affected Factory	NPD EC	Project maturity	Detailed EC type	Confirmed EC coordination status	External warehouse	Activity Count: Reminder CCB Buildability	Activity step order: CCB Buildability Confirmed
	0.23	0.09	0.08	0.06	0.04	0.03	0.02	0.02	0.01	0.01

Table 6.17 Top 10 most important features for XGB-CAGG by prefix length

Prefix length	Feature 1	Feature 2	Feature 3	Feature 4	Feature 5	Feature 6	Feature 7	Feature 8	Feature 9	Feature 10
1	Affected Model	NPD EC	Supplier Country	Affected Factory	Detailed EC type	External warehouse	Project maturity	Activity: Approval	Activity step order: purch. order complete	EC batch type
	0.25	0.15	0.11	0.10	0.04	0.03	0.02	0.02	0.01	0.01
2	Affected Model	NPD EC	Supplier Country	Affected Factory	Detailed EC type	External warehouse	Project maturity	Comm.ly relevant EC	Activity: Approval	Activity count: purch. order complete
	0.23	0.13	0.10	0.08	0.04	0.03	0.02	0.02	0.01	0.01
3	Affected Model	NPD EC	Supplier Country	Affected Factory	Detailed EC type	External warehouse	Project maturity	Activity step order: Check EC batch	Activity delay: Check EC	EC batch type
	0.18	0.10	0.10	0.08	0.03	0.02	0.02	0.02	0.01	0.01
4	Affected Model	NPD EC	Supplier Country	Affected Factory	External warehouse	Detailed EC type	Confirmed EC coordination status	Project maturity	Activity: Approval	Event date to first effectivity date (Check EC and EC batch)
	0.16	0.11	0.07	0.07	0.03	0.03	0.02	0.02	0.01	0.01

(continued)

Table 6.17 (continued)

Prefix length	Feature 1	Feature 2	Feature 3	Feature 4	Feature 5	Feature 6	Feature 7	Feature 8	Feature 9	Feature 10
5	Affected Model	Supplier Country	NPD EC	Affected Factory	Project maturity	External warehouse	Detailed EC type	EC scenario	Comm. relevant EC	Event date to planned purch. order date (Process Planning)
	0.15	0.11	0.06	0.05	0.03	0.02	0.02	0.01	0.01	0.01
6	Affected Model	NPD EC	Supplier Country	Affected Factory	Event date to effectivity date (SC confirmed)	Project maturity	Detailed EC type	External warehouse	Activity: Supply chain confirmed	Confirmed EC coordination status
	0.13	0.07	0.05	0.05	0.04	0.02	0.02	0.02	0.01	0.01
7	Affected Model	NPD EC	Supplier Country	Affected Factory	Activity Count: Supply Chain Confirmed	Detailed EC type	Project maturity	Event date to delivery date (Supply Chain confirmed)	Activity completion delay (CCB Buildability required)	Confirmed EC coordination status
	0.14	0.05	0.05	0.05	0.03	0.02	0.02	0.01	0.01	0.01
8	Affected Model	Event date to effectivity date (CCB Buildability required)	Activity Count: CCB Buildability Required	NPD EC	Supplier Country	Affected Factory	Event date to commercial approval deadline (CCB Buildability required)	Activity step order: Check EC batch	Event date to earliest possible Effectivity Date (Supply Chain Confirmed)	Event date to comm. approval deadline (Supply Chain Confirmed)
	0.10	0.06	0.06	0.06	0.04	0.04	0.03	0.02	0.02	0.02

(continued)

Table 6.17 (continued)

Prefix length	Feature 1	Feature 2	Feature 3	Feature 4	Feature 5	Feature 6	Feature 7	Feature 8	Feature 9	Feature 10
9	Affected Model	NPD EC	Activity Count: CCB Buildability Required	Event date to purchase order (CCB Buildability required)	Event date to effectivity date (CCB Buildability required)	Supplier Country	Affected Factory	Event date to delivery date (CCB Buildability required)	Detailed EC type	Event date to earliest possible Effectivity Date (Supply Chain Confirmed)
	0.08	0.06	0.06	0.06	0.05	0.04	0.04	0.02	0.02	0.01
10	Affected Model	Supplier Country	NPD EC	Affected Factory	Project maturity	External warehouse	Detailed EC type	Event date to delivery date (CCB confirmed)	Creation to first delivery	Event date to earliest effectivity date (ECN distributed)
	0.08	0.06	0.05	0.05	0.04	0.02	0.02	0.01	0.01	0.01
11	Affected Model	Supplier Country	Affected Factory	NPD EC	Project maturity	Event date to effectivity date (CCB Buildability confirmed)	Detailed EC type	Activity: CCB Buildability confirmed	Event date to purch. order deadline (Supply Chain Confirmed)	Event date to comm. approval deadline (SC confirmed)
	0.09	0.06	0.05	0.04	0.02	0.02	0.01	0.01	0.01	0.01
12	Supplier Country	Affected Model	NPD EC	Affected Factory	Project maturity	External warehouse	Detailed EC type	duration-T0062	Event date to purch. order deadline (Supply Chain Confirmed)	Activity: CCB Buildability confirmed
	0.08	0.08	0.06	0.04	0.02	0.02	0.01	0.01	0.01	0.01

(continued)

Table 6.17 (continued)

Prefix length	Feature 1	Feature 2	Feature 3	Feature 4	Feature 5	Feature 6	Feature 7	Feature 8	Feature 9	Feature 10
13	Event date to purchase order (CCB Buildability required)	Affected Model	Event date to effectivity date (CCB Buildability required)	NPD EC	Supplier Country	Affected Factory	Event date to comm. approval deadline (ECN complete)	Activity Count: CCB Buildability Required	Detailed EC type	Activity step order: Check EC batch
	0.11	0.08	0.08	0.07	0.04	0.04	0.02	0.02	0.01	0.01
14	Affected Model	NPD EC	Supplier Country	Project maturity	Affected Factory	External warehouse	Activity duration (Design solution)	Detailed EC type	Activity step order: Check EC batch	Event date to purch. order deadline (CCB confirmed)
	0.08	0.07	0.05	0.04	0.04	0.02	0.02	0.01	0.01	0.01
15	Affected Model	Supplier Country	NPD EC	Affected Factory	Project maturity	External warehouse	Event date to purch. order deadline (ECN complete)	Confirmed EC coordination status	Detailed EC type	Event date to earliest possible effectivity date (CCB Buildability)
	0.08	0.07	0.07	0.04	0.02	0.02	0.01	0.01	0.01	0.01

(continued)

Table 6.17 (continued)

Prefix length	Feature 1	Feature 2	Feature 3	Feature 4	Feature 5	Feature 6	Feature 7	Feature 8	Feature 9	Feature 10
16	Event date to effectivity date (CCB Buildability required)	Affected Model	Event date to purchase order (CCB Buildability required)	NPD EC	Supplier Country	Affected Factory	Activity duration (Comm. approval required)	Detailed EC type	Activity duration (Design solution)	External warehouse
	0.09	0.09	0.08	0.07	0.04	0.04	0.02	0.01	0.01	0.01
17	Event date to effectivity date (CCB Buildability required)	Event date to purchase order (CCB Buildability required)	Affected Model	NPD EC	Supplier Country	Affected Factory	Event date to purch. order deadline (Order availability)	Detailed EC type	External warehouse	Activity: Supply chain confirmed
	0.09	0.09	0.08	0.06	0.04	0.04	0.01	0.01	0.01	0.01
18	Event date to effectivity date (CCB Buildability required)	Event date to purchase order (CCB Buildability required)	Affected Model	NPD EC	Supplier Country	Affected Factory	Detailed EC type	Event date to purch. order deadline (Order availability)	Activity Count: CCB Buildability Required	External warehouse
	0.09	0.08	0.08	0.06	0.04	0.03	0.02	0.01	0.01	0.01
19	Event date to effectivity date (CCB Buildability required)	Affected Model	Event date to purchase order (CCB Buildability required)	NPD EC	Supplier Country	Affected Factory	Activity Count: CCB Buildability Required	Detailed EC type	External warehouse	Activity duration (Order availability)
	0.09	0.09	0.08	0.06	0.04	0.04	0.02	0.01	0.01	0.01

(continued)

Table 6.17 (continued)

Prefix length	Feature 1	Feature 2	Feature 3	Feature 4	Feature 5	Feature 6	Feature 7	Feature 8	Feature 9	Feature 10
20	Affected Model	Event date to effectivity date (CCB Buildability required)	NPD EC	Event date to purchase order (CCB Buildability required)	Supplier Country	Affected Factory	Activity duration (Comm. approval required)	Detailed EC type	Activity Count: CCB Buildability Required	External warehouse
	0.08	0.08	0.07	0.06	0.05	0.04	0.03	0.02	0.01	0.01
21	Affected Model	NPD EC	Supplier Country	Affected Factory	Project maturity	External warehouse	Activity duration (Design solution)	Event date to purch. order deadline (CCB confirmed)	Detailed EC type	Creation to first delivery
	0.07	0.06	0.05	0.04	0.04	0.02	0.01	0.01	0.01	0.01
22	Event date to purchase order (CCB Buildability required)	Event date to effectivity date (CCB Buildability required)	Affected Model	NPD EC	Supplier Country	Affected Factory	Detailed EC type	Activity count: Comm. approval required	External warehouse	Confirmed EC coordination status
	0.12	0.11	0.08	0.06	0.04	0.03	0.02	0.01	0.01	0.01
23	Event date to effectivity date (CCB Buildability required)	Event date to purchase order (CCB Buildability required)	Affected Model	NPD EC	Supplier Country	Activity Count: CCB Buildability Required	Affected Factory	Detailed EC type	External warehouse	Confirmed EC coordination status
	0.12	0.08	0.08	0.06	0.04	0.04	0.03	0.02	0.01	0.01

(continued)

Table 6.17 (continued)

Prefix length	Feature 1	Feature 2	Feature 3	Feature 4	Feature 5	Feature 6	Feature 7	Feature 8	Feature 9	Feature 10
24	Affected Model	Supplier Country	Affected Factory	NPD EC	Event date to purchase order (CCB Buildability required)	Event date to purchase order (Check EC and EC batch)	Event date to delivery date (Check EC and EC batch)	Project maturity	Detailed EC type	Activity: CCB Buildability required
	0.10	0.05	0.04	0.04	0.03	0.02	0.02	0.01	0.01	0.01
25	Affected Model	Supplier Country	Affected Factory	NPD EC	Project maturity	Activity: CCB Buildability required	Event date to purch. order deadline (ECN complete)	Detailed EC type	Confirmed EC coordination status	Activity duration (CCB Buildability required)
	0.08	0.05	0.04	0.03	0.02	0.02	0.01	0.01	0.01	0.01
26	Affected Model	Supplier Country	Affected Factory	NPD EC	Activity: CCB Buildability required	Project maturity	Detailed EC type	Activity: CCB Buildability confirmed	Event date to earliest possible effectivity date (CCB Buildability)	Confirmed EC coordination status
	0.08	0.06	0.04	0.04	0.02	0.02	0.01	0.01	0.01	0.01

(continued)

Table 6.17 (continued)

Prefix length	Feature 1	Feature 2	Feature 3	Feature 4	Feature 5	Feature 6	Feature 7	Feature 8	Feature 9	Feature 10
27	Affected Model	Supplier Country	Affected Factory	Activity: CCB Buildability required	NPD EC	Project maturity	Detailed EC type	Activity completion delay (Reminder CCB Buildability)	Confirmed EC coordination status	Event date to earliest possible effectivity date (CCB Buildability)
	0.08	0.06	0.04	0.04	0.04	0.02	0.01	0.01	0.01	0.01
28	Affected Model	Supplier Country	Affected Factory	NPD EC	Activity: CCB Buildability required	Project maturity	Activity completion delay (Reminder CCB Buildability)	Event date to earliest possible effectivity date (CCB Buildability)	Detailed EC type	Event date to earliest possible effectivity date (CCB Buildability)
	0.08	0.05	0.04	0.03	0.03	0.02	0.01	0.01	0.01	0.01
29	Affected Model	Supplier Country	Activity: CCB Buildability required	Affected Factory	NPD EC	Project maturity	Activity completion delay (Reminder CCB Buildability)	Detailed EC type	Confirmed EC coordination status	Event date to earliest possible effectivity date (CCB Buildability)
	0.08	0.05	0.04	0.04	0.03	0.02	0.01	0.01	0.01	0.01

Table 6.18 Top 10 most important features for XGB-IDX by prefix length

Prefix length	Feature 1	Feature 2	Feature 3	Feature 4	Feature 5	Feature 6	Feature 7	Feature 8	Feature 9	Feature 10
1	Affected Model	NPD EC	Affected Factory	Supplier Country	Detailed EC type	Project maturity	External warehouse	EC batch type	EC scenario	EC process type
	0.27	0.14	0.12	0.11	0.05	0.03	0.03	0.01	0.01	0.01
2	Affected Model	Supplier Country	NPD EC	Affected Factory	Detailed EC type	External warehouse	Project maturity	EC scenario	EC batch type	EC process type
	0.26	0.12	0.11	0.09	0.05	0.03	0.03	0.01	0.01	0.01
3	Affected Model	NPD EC	Supplier Country	Affected Factory	Detailed EC type	External warehouse	Project maturity	1st activity description: Check EC and EC batch	Event date to purch. order deadline (2nd activity)	Activity has target date (2nd Activity)
	0.20	0.17	0.10	0.08	0.03	0.03	0.02	0.01	0.01	0.01
4	Affected Model	NPD EC	Supplier Country	Affected Factory	Detailed EC type	External warehouse	Project maturity	Confirmed EC coordination status	Activity has target date (2nd Activity)	Event date to creation (1st activity)
	0.19	0.10	0.09	0.07	0.05	0.03	0.03	0.01	0.01	0.01

(continued)

Table 6.18 (continued)

Prefix length	Feature 1	Feature 2	Feature 3	Feature 4	Feature 5	Feature 6	Feature 7	Feature 8	Feature 9	Feature 10
5	Affected Model	NPD EC	Supplier Country	Affected Factory	Detailed EC type	Project maturity	External warehouse	EC scenario	Event date to comm. approval deadline (3rd activity)	2nd activity description: Check EC and EC batch
	0.18	0.09	0.09	0.08	0.03	0.03	0.02	0.01	0.01	0.01
6	Affected Model	Supplier Country	NPD EC	Affected Factory	Project maturity	External warehouse	Detailed EC type	4th activity description: Check part number	Confirmed EC coordination status	Commercially relevant EC
	0.15	0.08	0.07	0.06	0.03	0.03	0.02	0.01	0.01	0.01
7	Affected Model	Supplier Country	NPD EC	Affected Factory	Project maturity	External warehouse	Detailed EC type	Event date to creation (6th activity)	Confirmed EC coordination status	Creation to effectivity date
	0.11	0.07	0.07	0.05	0.04	0.02	0.02	0.01	0.01	0.01
8	Affected Model	Supplier Country	Affected Factory	NPD EC	7th activity description: Check part number	Project maturity	Detailed EC type	External warehouse	6th activity description: Confirm supply chain	EC scenario
	0.12	0.09	0.04	0.04	0.03	0.02	0.02	0.02	0.01	0.01

(continued)

Table 6.18 (continued)

Prefix length	Feature 1	Feature 2	Feature 3	Feature 4	Feature 5	Feature 6	Feature 7	Feature 8	Feature 9	Feature 10
9	Affected Model	Supplier Country	NPD EC	Affected Factory	Project maturity	External warehouse	Event date to creation (6th activity)	Detailed EC type	Activity has target date (7th activity)	Event date to first effectivity date (2nd activity)
	0.10	0.06	0.06	0.04	0.04	0.03	0.02	0.01	0.01	0.01
10	Affected Model	NPD EC	Supplier Country	Affected Factory	Project maturity	External warehouse	Detailed EC type	Event date to creation (6th activity)	4th activity description: Check part number	Confirmed EC coordination status
	0.08	0.07	0.06	0.04	0.03	0.02	0.01	0.01	0.01	0.01
11	Affected Model	NPD EC	Supplier Country	Project maturity	Affected Factory	External warehouse	Detailed EC type	Event date to creation (6th activity)	Confirmed EC coordination status	Event date to first effectivity date (2nd activity)
	0.08	0.06	0.06	0.03	0.03	0.02	0.01	0.01	0.01	0.01
12	Affected Model	NPD EC	Supplier Country	Affected Factory	Project maturity	External warehouse	Event date to creation (6th activity)	Detailed EC type	Event date to first effectivity date (2nd activity)	Event date to delivery date (2nd activity)
	0.08	0.06	0.05	0.04	0.03	0.02	0.01	0.01	0.01	0.01

(continued)

Table 6.18 (continued)

Prefix length	Feature 1	Feature 2	Feature 3	Feature 4	Feature 5	Feature 6	Feature 7	Feature 8	Feature 9	Feature 10
13	Affected Model	NPD EC	Supplier Country	Affected Factory	Project maturity	External warehouse	Event date to creation (6th activity)	Detailed EC type	Event date to delivery date (2nd activity)	Event date to first effectivity date (2nd activity)
	0.07	0.05	0.05	0.04	0.03	0.02	0.01	0.01	0.01	0.01
14	Affected Model	NPD EC	Supplier Country	Affected Factory	Project maturity	External warehouse	Detailed EC type	Event date to creation (6th activity)	Confirmed EC coordination status	Creation to effectivity date
	0.06	0.05	0.05	0.03	0.03	0.02	0.01	0.01	0.01	0.01
15	Affected Model	NPD EC	Supplier Country	Affected Factory	Project maturity	External warehouse	Detailed EC type	Event date to creation (1st activity)	Event date to creation (6th activity)	Event date to purch. order deadline (1st activity)
	0.06	0.04	0.04	0.03	0.03	0.02	0.01	0.01	0.01	0.01
16	Affected Model	NPD EC	Supplier Country	Affected Factory	Project maturity	External warehouse	Event date to purch. order deadline (1st activity)	Detailed EC type	Event date to creation (6th activity)	4th activity description: Check part number
	0.06	0.04	0.04	0.03	0.03	0.02	0.01	0.01	0.01	0.01

(continued)

Table 6.18 (continued)

Prefix length	Feature 1	Feature 2	Feature 3	Feature 4	Feature 5	Feature 6	Feature 7	Feature 8	Feature 9	Feature 10
17	Affected Model	NPD EC	Supplier Country	Affected Factory	Project maturity	External warehouse	Event date to creation (1st activity)	Detailed EC type	Confirmed EC coordination status	Event date to purch. order deadline (1st activity)
	0.06	0.04	0.04	0.03	0.03	0.01	0.01	0.01	0.01	0.01
18	Affected Model	NPD EC	Supplier Country	Affected Factory	Project maturity	External warehouse	Event date to creation (6th activity)	Event date to creation (1st activity)	Event date to purch. order deadline (1st activity)	Detailed EC type
	0.06	0.04	0.03	0.03	0.03	0.02	0.01	0.01	0.01	0.01
19	Affected Model	NPD EC	Supplier Country	Affected Factory	Project maturity	External warehouse	Detailed EC type	Event date to creation (6th activity)	Event date to purch. order deadline (1st activity)	Event date to creation (1st activity)
	0.05	0.04	0.03	0.03	0.03	0.01	0.01	0.01	0.01	0.01
20	Affected Model	NPD EC	Supplier Country	Affected Factory	Project maturity	External warehouse	Event date to creation (6th activity)	Event date to purch. order deadline (1st activity)	Detailed EC type	Event date to creation (1st activity)
	0.05	0.04	0.03	0.03	0.03	0.01	0.01	0.01	0.01	0.01

(continued)

Table 6.18 (continued)

Prefix length	Feature 1	Feature 2	Feature 3	Feature 4	Feature 5	Feature 6	Feature 7	Feature 8	Feature 9	Feature 10
21	Affected Model	NPD EC	Supplier Country	Project maturity	Affected Factory	External warehouse	Event date to creation (6th activity)	Detailed EC type	Event date to creation (1st activity)	Event date to purch. order deadline (1st activity)
	0.05	0.04	0.03	0.03	0.03	0.01	0.01	0.01	0.01	0.01
22	Affected Model	NPD EC	Supplier Country	Affected Factory	Project maturity	External warehouse	Event date to creation (6th activity)	Event date to purch. order deadline (1st activity)	Detailed EC type	Event date to creation (1st activity)
	0.04	0.04	0.03	0.03	0.02	0.01	0.01	0.01	0.01	0.01
23	Affected Model	NPD EC	Supplier Country	Affected Factory	Project maturity	External warehouse	Confirmed EC coordination status	Event date to creation (1st activity)	Event date to purch. order deadline (1st activity)	Event date to creation (6th activity)
	0.05	0.04	0.03	0.03	0.02	0.01	0.01	0.01	0.01	0.01
24	Affected Model	NPD EC	Supplier Country	Affected Factory	Project maturity	External warehouse	Event date to creation (1st activity)	Confirmed EC coordination status	Detailed EC type	Event date to creation (6th activity)
	0.05	0.03	0.03	0.03	0.02	0.01	0.01	0.01	0.01	0.01

(continued)

Table 6.18 (continued)

Prefix length	Feature 1	Feature 2	Feature 3	Feature 4	Feature 5	Feature 6	Feature 7	Feature 8	Feature 9	Feature 10
25	Affected Model	NPD EC	Supplier Country	Affected Factory	Project maturity	External warehouse	Detailed EC type	Event date to purch. order deadline (1st activity)	Confirmed EC coordination status	Event date to creation (6th activity)
	0.05	0.04	0.03	0.02	0.02	0.01	0.01	0.01	0.01	0.01
26	Affected Model	NPD EC	Supplier Country	Affected Factory	Project maturity	Detailed EC type	External warehouse	Confirmed EC coordination status	Event date to creation (1st activity)	Event date to creation (6th activity)
	0.06	0.04	0.03	0.02	0.02	0.01	0.01	0.01	0.01	0.01
27	Affected Model	NPD EC	Supplier Country	Affected Factory	Project maturity	External warehouse	Detailed EC type	Event date to creation (6th activity)	Event date to purch. order deadline (1st activity)	Confirmed EC coordination status
	0.04	0.04	0.03	0.03	0.02	0.01	0.01	0.01	0.01	0.01

(continued)

Table 6.18 (continued)

Prefix length	Feature 1	Feature 2	Feature 3	Feature 4	Feature 5	Feature 6	Feature 7	Feature 8	Feature 9	Feature 10
28	Affected Model	NPD EC	Supplier Country	Affected Factory	Project maturity	External warehouse	Event date to creation (6th activity)	Confirmed EC coordination status	Detailed EC type	Event date to comm. approval deadline (2nd activity)5
	0.04	0.03	0.03	0.03	0.02	0.01	0.01	0.01	0.01	0.01
29	Affected Model	NPD EC	Supplier Country	Affected Factory	Project maturity	External warehouse	Event date to creation (6th activity)	Detailed EC type	Event date to creation (1st activity)	Event date to purch. order deadline (1st activity)
	0.04	0.04	0.03	0.02	0.02	0.01	0.01	0.01	0.01	0.01

6.9 Presentation of Own Contribution

All data presented within the publication was collected and the contents processed by me. This includes the definition of the research scope and research questions, screening of theoretical background research, identification of relevant data bases, definition of test sets, identification and application of the learning algorithms, and definition of the evaluation criteria. Furthermore, the implementation of the learning algorithms into Python code, the development of additional encoding methods, the execution of the test code, and the evaluation of the results was done by me. Additionally, the textual elaboration and writing of the publication were entirely conducted by me. All four co-authors Prof. Dr. Burggräf, Dr. Steinberg, Mr. Becher, and Dr. Weißer contributed with ideas to the research concept and in the internal review process.

7 Critical Reflection and Future Perspective

The four main publications composing the cumulative part of this thesis describe the application of AI-methods within ECM. Following the systematic literature review in Chap. 3, the other three publications are each dedicated to one of the business needs described in Chap. 2, showing great promise for AI-application as parts of an automatic ECM system. However, the methods and algorithms were all applied within the confined boundaries of a single automotive OEM. As commonly experienced within EC research, data is only available as simulated cases or with limited accessibility. Furthermore, the data was based on an automotive use-case. EC, however, is common within multiple industries. Moreso, it has been shown that engineer-to-order companies experience different difficulties to those seen in mass producing ones. Consequently, the performance of those approaches presented in the publications II, III, and IV (cf. Chaps. 4–6) should be evaluated both in other companies and industries.

To further build on the data aspect, within the publications, the ECM system was considered as the main information source for prediction. However, as seen in publication III (cf. Chap. 5), quality induced EC processes are not well described by it. Therefore, it would be relevant to check how data from complementary source systems, e.g., problem management systems, impacts predictive performance. Subsequently, in publication III and IV only structured data has been used. As EC data is partly also available as textual information, it should be investigated how it could be incorporated and applied as input data to improve predictive performance.

In addition, the approaches have each only applied basic algorithms. As recent research has shown, advanced algorithms or combinations of different basic algorithms have the potential to improve performance with little effort. Thus, it would

O. Radišić-Aberger, *Optimisation and Control of Engineering Change Schedules in the Automotive Industry with Metaheuristics and Machine Learning*, Findings from Production Management Research,
https://doi.org/10.1007/978-3-658-51074-9_7

be interesting to test whether certain combinations of ML and MH algorithms could improve the optimisation and prediction performance.

Finally, for the goal of an autonomous EC scheduling system, research in combinations of the different approaches should be conducted. Within the developed approaches, a human operator was still required at the centre to trigger each of the calculations. However, to draw the maximum benefit from AI, a fully autonomous operational system should be devised. Given sufficient information, this research could see the introduction of reinforcement learning to the EC domain. A starting point could be transforming the predictive nature of the "Supervisor" agent into a prescriptive method, thereby aiding human operators in laying out ideal process flows. Building on that, a reinforcement learning approach could learn from various prescribed processes which ones are actually feasible. Ultimately, the reward metric would be the information whether a human had to adjust the schedule and whether the costs were acceptable.

Summary

8

Within this thesis AI-algorithms were applied to optimise, predict, and control the EC implementation process, thus laying the foundation of a possible automatic EC management and scheduling system. Based on an initial literature search, a need for computational support in EC implementation has been established, described by four BuNe: optimisation, decision support, supervision, and automation.

To support their fulfilment and solution development, a knowledge base of AI-artefacts in ECM was established based on a systematic literature review. Subsequently, it was presented that AI-methods are applicable in the domain, although mainly focusing on earlier phases within the EC process. Of the used methods, MH algorithms and ML models are the most widespread AI-applications.

Starting with the first BuNe, a benchmarking approach has been applied on a newly developed mathematical model to identify the most suitable MH for optimising the EC effectivity date as a function of stock and quality costs. Calculating with real-world data from an industry partner, the GA has provided the best results at the highest reliability. However, due to long computing times, it was determined that a sequential usage of GA and TS performs best at determining EC effectivity dates. The resulting IT-tool, validated at the industry partner, has thus shown how a combined MH approach can efficiently determine and control EC implementation costs.

Having established a method to calculate an optimal EC effectivity date, its feasibility had to be evaluated. Therefore, ML models were tested to see whether it is possible to predict EC schedule adherence three weeks in advance. To decide whether the performance is sensible from a business perspective, four supervised

O. Radišić-Aberger, *Optimisation and Control of Engineering Change Schedules in the Automotive Industry with Metaheuristics and Machine Learning*, Findings from Production Management Research,
https://doi.org/10.1007/978-3-658-51074-9_8

learning algorithms were trained on past real-world EC data from an industry partner. As the research has shown, the shallow learning approaches can predict the adherence to EC effectivity dates. Integrating the trained models into a web-tool, it was further shown how an ML based artefact can be applied for determining the plausibility of EC effectivity dates.

Once it was established that adherence to the EC effectivity date can be predicted, it was investigated how ML models perform at earlier process steps, to fulfil the third BuNe, 'supervision'. Therefore, using PBPM methods, a set of ML algorithm-encoding combinations was trained with data varying in process length and applied to the industry-provided EC test set. Demonstrating that early predictive performance of ML models is nearly equal to those in later EC process steps, it is shown how ML models integrated into an EC monitoring tool provide necessary decision support for EC control across all EC phases.

Overall, this work exhibits applications of AI methods for improving the EC implementation phase. The displayed methods showcase how to solve underlying challenges, such as optimisation and control, as well as how to evaluate developed methods. The three publications discussing the agents provide a baseline for future research and can be applied as standalone programs coordinated by a human operator. In total, they form the core of future automated EC implementation.

References

1. O. Radisic-Aberger, T. Weisser, T. Saßmannshausen, J. Wagner, and P. Burggräf, “Concept of a Multi-Agent System for Optimised and Automated Engineering Change Implementation,” *Proceedings of the Design Society*, vol. 2, pp. 1689–1698, 2022, https://doi.org/10.1017/pds.2022.171.
2. G. Schuh, J.-P. Prote, M. Luckert, F. Basse, V. Thomson, and W. Mazurek, “Adaptive Design of Engineering Change Management in Highly Iterative Product Development,” *Procedia CIRP*, vol. 70, pp. 72–77, 2018, https://doi.org/10.1016/j.procir.2018.02.016.
3. G. Schuh and M. Riesener, *Produktkomplexität managen*. München: Carl Hanser Verlag GmbH & Co. KG, 2017.
4. B. Hamraz, N. H. M. Caldwell, and P. J. Clarkson, “A Holistic Categorization Framework for Literature on Engineering Change Management,” *Syst. Engin.*, vol. 16, no. 4, pp. 473–505, 2013, https://doi.org/10.1002/sys.21244.
5. F. B. Watts, “Change Management,” in *Configuration management for senior managers: Essential product configuration and lifecycle management for manufacturing*, F. B. Watts, Ed., Kidlington, Oxford, UK: Butterworth Heinemann is an imprint of Elsevier, 2015, pp. 123–148.
6. N. Iakymenko, P. O. Brett, E. Alfnes, and J. O. Strandhagen, “Analyzing the factors affecting engineering change implementation performance in the engineer-to-order production environment: case studies from a Norwegian shipbuilding group,” *Production Planning & Control*, pp. 1–17, 2020, https://doi.org/10.1080/09537287.2020.1837939.
7. R. Capistrano Burgos, F. Sippl, O. Radisic-Aberger, and T. Weisser, “Data-Based Method for the Implementation Planning of Engineering Changes in the Automotive Industry,” *Proceedings of the Design Society*, vol. 2, pp. 343–352, 2022, https://doi.org/10.1017/pds.2022.36.
8. M. A. Fletcher, “General Motors recalls another 2.7 million vehicles,” *The Washington Post*, 15 May., 2014. https://www.washingtonpost.com/business/economy/general-motors-recalls-another-27-million-vehicles/2014/05/15/f8bce616-dc40-11e3-bda1-9b46b2066796_story.html

O. Radišić-Aberger, *Optimisation and Control of Engineering Change Schedules in the Automotive Industry with Metaheuristics and Machine Learning*, Findings from Production Management Research, https://doi.org/10.1007/978-3-658-51074-9

9. B. Vlasic, "An Engineer's Eureka Moment With a G.M. Flaw," *The New York Times*, 28 Mar., 2014. https://www.nytimes.com/2014/03/29/business/a-florida-engineer-unlocked-the-mystery-of-gms-ignition-flaw.html
10. M. C. Wickel, "Änderungen besser managen: Eine datenbasierte Methodik zur Analyse technischer Änderungen," Dissertation, Technische Universität München, München, 2017. [Online]. Available: https://mediatum.ub.tum.de/?id=1310063
11. A. Wasmer, G. Staub, and R. W. Vroom, "An industry approach to shared, cross-organisational engineering change handling - The road towards standards for product data processing," *Computer-Aided Design*, vol. 43, no. 5, pp. 533–545, 2011, https://doi.org/10.1016/j.cad.2010.10.002.
12. E. Fricke, B. Gebhard, H. Negele, and E. Igenbergs, "Coping with changes: Causes, findings, and strategies," *Systems Engineering*, vol. 3, no. 4, pp. 169–179, 2000, https://doi.org/10.1002/1520-6858(2000)3:4<169::AID-SYS1>3.0.CO;2-W.
13. P. J. Clarkson, C. Simons, and C. Eckert, "Predicting Change Propagation in Complex Design," *Journal of Mechanical Design*, vol. 126, no. 5, pp. 788–797, 2004, https://doi.org/10.1115/1.1765117.
14. C. Eckert, P. J. Clarkson, and W. Zanker, "Change and customisation in complex engineering domains," *Research in Engineering Design*, vol. 15, no. 1, pp. 1–21, 2004, https://doi.org/10.1007/s00163-003-0031-7.
15. G. Schuh, S. Aleksic, and S. Rudolf, "Module-based release management for technical changes," in *Advances in Intelligent Systems and Computing, Progress in Systems Engineering*, H. Selvaraj, D. Zydek, and G. Chmaj, Eds., Cham: Springer International Publishing, 2015, pp. 293–298.
16. A. Brahma and D. C. Wynn, "Concepts of change propagation analysis in engineering design," *Research in Engineering Design*, 2022, https://doi.org/10.1007/s00163-022-00395-y.
17. C. Wänström and P. Jonsson, "The impact of engineering changes on materials planning," *Journal of Manufacturing Technology Management*, vol. 17, no. 5, pp. 561–584, 2006, https://doi.org/10.1108/17410380610668522.
18. S. Liu, X. Meng, and B. Gong, "Modeling and implementing of a flexible workflow system which supporting engineering change," in *8th International Conference on Computer Supported Cooperative Work in Design*, Xiamen, China, 2004, pp. 418–422.
19. W. R. Ashby, *An introduction to cybernetics*. Mansfield Centre, CT: Martino Publishing, 2015.
20. M. E. Sharp, T. D. Hedberg, W. Z. Bernstein, and S. Kwon, "Feasibility study for an automated engineering change process," *International Journal of Production Research*, vol. 59, no. 16, pp. 4995–5010, 2021, https://doi.org/10.1080/00207543.2021.1893900.
21. P. Zheng, C.-H. Chen, and S. Shang, "Towards an automatic engineering change management in smart product-service systems – A DSM-based learning approach," *Advanced Engineering Informatics*, vol. 39, pp. 203–213, 2019, https://doi.org/10.1016/j.aei.2019.01.002.
22. D. S. Shivankar and R. Deivanathan, "Product design change propagation in automotive supply chain considering product life cycle," *CIRP Journal of Manufacturing Science and Technology*, vol. 35, pp. 390–399, 2021, https://doi.org/10.1016/j.cirpj.2021.07.001.

23. R. Barzizza, M. Caridi, and R. Cigolini, "Engineering change: A theoretical assessment and a case study," *Production Planning & Control*, vol. 12, no. 7, pp. 717–726, 2001, https://doi.org/10.1080/09537280010024054.
24. J.-Y. Shiau, "Effectivity date analysis and scheduling," *International Journal of Production Research*, vol. 49, no. 10, pp. 2771–2791, 2011, https://doi.org/10.1080/00207541003713017.
25. H. A. Taha, *Operations Research: An Introduction, Global Edition,* 10th ed. Harlow: Pearson Education Limited, 2017. [Online]. Available: https://elibrary.pearson.de/book/99.150005/9781292165561
26. C. Wänström, F. Lind, and O. Wintertidh, "Creating a model to facilitate the allocation of materials planning resources in engineering change situations," *International Journal of Production Research*, vol. 44, 18–19, pp. 3775–3796, 2006, https://doi.org/10.1080/00207540600622506.
27. A. Stekolschik, "Engineering Change Management Method Framework in Mechanical Engineering," *IOP Conference Series: Materials Science and Engineering*, vol. 157, p. 12008, 2016, https://doi.org/10.1088/1757-899X/157/1/012008.
28. C. Di Francescomarino and C. Ghidini, "Predictive Process Monitoring," in *Process Mining Handbook*: Springer, Cham, 2022, pp. 320–346. [Online]. Available: https://link.springer.com/chapter/10.1007/978-3-031-08848-3_10
29. M. Dumas, M. La Rosa, J. Mendling, and H. A. Reijers, *Fundamentals of Business Process Management*. Berlin, Heidelberg: Springer Berlin Heidelberg, 2018.
30. M. Wickel, N. Chucholowski, F. Behncke, and U. Lindemann, "Comparison of Seven Company-Specific Engineering Change Processes," in *Modelling and Management of Engineering Processes*, M. Schabacker, K. Gericke, N. Szélig, and S. Vajna, Eds., Berlin, Heidelberg: Springer Berlin Heidelberg, 2015, pp. 125–136.
31. L. Maruster and A. Alblas, "Tailoring the Engineering Design Process Through Data and Process Mining," *IEEE Trans. Eng. Manage.*, vol. 69, no. 4, pp. 1577–1591, 2022, https://doi.org/10.1109/TEM.2020.3000861.
32. O. Radišić-Aberger, P. Burggräf, F. Steinberg, A. Becher, and T. Weißer, "Evaluating early predictive performance of machine learning approaches for engineering change schedule – A case study using predictive process monitoring techniques," *Supply Chain Analytics*, vol. 8, p. 100087, 2024, https://doi.org/10.1016/j.sca.2024.100087.
33. N. Kattner, J. Mehlstaeubl, L. Becerril, and U. Lindemann, "Data Analysis in Engineering Change Management—Improving Collaboration by Assessing Organizational Dependencies Based on Past Engineering Change Information," in *2018 IEEE International Conference on Industrial Engineering and Engineering Management (IEEM)*, Bangkok, 2018, pp. 617–621.
34. S. J. Russell and P. Norvig, *Artificial intelligence: A modern approach*. Boston, Columbus, Indianapolis: Pearson, 2016. [Online]. Available: https://ebookcentral.proquest.com/lib/kxp/detail.action?docID=5831883
35. I. Teinemaa, M. Dumas, M. La Rosa, and F. M. Maggi, "Outcome-Oriented Predictive Process Monitoring," *ACM Trans. Knowl. Discov. Data*, vol. 13, no. 2, pp. 1–57, 2019, https://doi.org/10.1145/3301300.
36. N. J. Nilsson, *The quest for artificial intelligence: A history of ideas and achievements.* Cambridge: Cambridge Univ. Press, 2010.
37. R. Carnap, "Formalwissenschaft und Realwissenschaft," *Erkenntnis*, vol. 5, no. 1, pp. 30–37, 1935, https://doi.org/10.1007/BF00172279.

38. Hevner, March, Park, and Ram, "Design Science in Information Systems Research," *MIS Quarterly*, vol. 28, no. 1, p. 75, 2004, https://doi.org/10.2307/25148625.
39. K. Peffers, T. Tuunanen, and B. Niehaves, "Design science research genres: introduction to the special issue on exemplars and criteria for applicable design science research," *European Journal of Information Systems*, vol. 27, no. 2, pp. 129–139, 2018, https://doi.org/10.1080/0960085X.2018.1458066.
40. C. M. Eckert, R. Keller, and P. J. Clarkson, "Change prediction in innovative products to avoid emergency innovation," *International Journal of Technology Management*, vol. 55, ¾, p. 226, 2011, https://doi.org/10.1504/IJTM.2011.041949.
41. W. Wei, "Study of claim support system for hydraulic engineering based on CBR and RBR," in *2010 5th International Conference on Computer Sciences and Convergence Information Technology (ICCIT 2010): Seoul, South Korea, 30 November - 2 December 2010 ; [proceeding*, Seoul, 2010, pp. 763–765.
42. V. N. Kravets, J.-H. R. Jiang, and H. Riener, "Learning to Automate the Design Updates From Observed Engineering Changes in the Chip Development Cycle," in *Proceedings of the 2020 Design, Automation & Test in Europe Conference & Exhibition (DATE 2020): 09 to 13 March, 2020, Grenoble, France*, Grenoble, France, 2020, pp. 738–743.
43. T. A. W. Jarratt, C. M. Eckert, N. H. M. Caldwell, and P. J. Clarkson, "Engineering change: an overview and perspective on the literature," *Research in Engineering Design*, vol. 22, no. 2, pp. 103–124, 2011a, https://doi.org/10.1007/s00163-010-0097-y.
44. P. Burggräf, J. Wagner, T. Saßmannshausen, T. Weißer, and O. Radisic-Aberger, "AI-artifacts in engineering change management—a systematic literature review," *Research in Engineering Design*, 2024a, https://doi.org/10.1007/s00163-023-00430-6.
45. T. A. W. Jarratt, "A model-based approach to support the management of engineering change," Apollo—University of Cambridge Repository, 2004.
46. T. Jarratt, J. Clarkson, and C. Eckert, "Engineering change," in *Design process improvement: A review of current practice*, J. Clarkson and C. Eckert, Eds., London: Springer, 2005b, pp. 262–285.
47. E. Monahan, *Engineering Documentation Control Practices and Procedures*. Boca Raton: Chapman and Hall/CRC, 1995. [Online]. Available: https://ebookcentral.proquest.com/lib/gbv/detail.action?docID=5379057
48. S. Prakash and J. Schlotthauer, "Simulation of engineering design change approval-process," in *Proceedings of the 24th conference on Winter simulation - WSC '92*, Arlington, Virginia, United States, 1992, pp. 842–848.
49. G. Q. Huang and K. L. Mak, "Current practices of engineering change management in UK manufacturing industries," *Int Jrnl of Op & Prod Mnagemnt*, vol. 19, no. 1, pp. 21–37, 1999, https://doi.org/10.1108/01443579910244205.
50. A. Riviere, C. DaCunha, and M. Tollenaere, "Performances in Engineering Changes Management," in *Recent Advances in Integrated Design and Manufacturing in Mechanical Engineering*, G. Gogu, D. Coutellier, P. Chedmail, and P. Ray, Eds., Dordrecht: Springer, 2003, pp. 369–378.
51. N. Balakrishnan and A. K. Chakravarty, "Managing Engineering Change: Market Opportunities and Manufacturing Costs," *Production and Operations Management*, vol. 5, no. 4, pp. 335–356, 1996, https://doi.org/10.1111/j.1937-5956.1996.tb00404.x.

52. N. Iakymenko, A. Romsdal, E. Alfnes, M. Semini, and J. O. Strandhagen, "Status of engineering change management in the engineer-to-order production environment: insights from a multiple case study," *International Journal of Production Research*, vol. 58, no. 15, pp. 4506–4528, 2020, https://doi.org/10.1080/00207543.2020.1759836.
53. M. Z. Ouertani and K. Grebici, "Supporting conflict management in collaborative design: An approach to assess engineering change impacts," *Computers in Industry*, vol. 59, no. 9, pp. 882–893, 2008, https://doi.org/10.1016/j.compind.2010.08.001.
54. P. Potdar and V. Jonnalagedda, "Design and development of a framework for effective engineering change management in manufacturing industries," *IJPLM*, vol. 11, no. 4, p. 368, 2018, https://doi.org/10.1504/ijplm.2018.097880.
55. K. J. Balcerak and B. G. Dale, "Engineering change administration: the key issues," *Computer Integrated Manufacturing Systems*, vol. 5, no. 2, pp. 125–132, 1992, https://doi.org/10.1016/0951-5240(92)90007-Y.
56. M. Diprima, "Engineering Change Control and Implementation Consideration," *Production & Inventory Management*, vol. 23, no. 1, pp. 81–87, 1982.
57. M. A. Reidelbach, "Engineering Change Management for Long-Lead Time Production Environments," *Production & Inventory Management Journal*, vol. 32, no. 2, pp. 84–88, 1991.
58. N. Iakymenko, A. Romsdal, M. Semini, and J. O. Strandhagen, "Managing engineering changes in the engineer-to-order environment: challenges and research needs," *IFAC-PapersOnLine*, vol. 51, no. 11, pp. 144–151, 2018, https://doi.org/10.1016/j.ifacol.2018.08.249.
59. L. H. Li, R. Mo, H. B. Sun, and Z. Y. Chang, "Petri Nets-Based Implement Process Model for Engineering Change on Aero-Engine Assembly Field," *AMM*, 148–149, pp. 1118–1121, 2011a, https://doi.org/10.4028/www.scientific.net/AMM.148-149.1118.
60. W. Li and Y. B. Moon, "A Simulation Model for Managing Engineering Changes Along with New Product Development," in *Proceedings of the 11th International Conference on Enterprise Information*, Milan, Italy, 2009a, pp. 13–18.
61. W. Li and Y. B. Moon, "A simulation study of mutual influences of Engineering Change Management process and New Product Development process," in *Proceedings of the 2009 Winter Simulation Conference (WSC)*, Austin, TX, USA, 2009, pp. 2940–2950.
62. W. Li and Y. B. Moon, "Modeling and managing Engineering Changes in a complex product development process," in *Proceedings of the 2011 Winter Simulation Conference (WSC)*, Phoenix, AZ, USA, 2011b, pp. 792–804.
63. N. Bhuiyan, G. Gatard, and V. Thomson, "Engineering change request management in a new product development process," *Euro Jrnl of Inn Mnagmnt*, vol. 9, no. 1, pp. 5–19, 2006, https://doi.org/10.1108/14601060610639999.
64. N. Ahmad, D. C. Wynn, and P. J. Clarkson, "The Impact of Packaging Interdependent Change Requests on ProjectLead Time," *DSM 2010: Proceedings of the 12th International DSM Conference, Cambridge, UK, 22.-23.07.2010*, pp. 293–306, 2010.
65. D. C. Wynn, N. H. M. Caldwell, and P. J. Clarkson, "Can Change Prediction help Prioritise Redesign Work in Future Engineering Systems?," *DS 60: Proceedings of DESIGN 2010, the 11th International Design Conference, Dubrovnik, Croatia*, pp. 1691–1702, 2010a.

66. M. Z. Ouertani and L. Gzara, "Tracking product specification dependencies in collaborative design for conflict management," *Computer-Aided Design*, vol. 40, no. 7, pp. 828–837, 2008, https://doi.org/10.1016/j.cad.2007.07.002.
67. M. Z. Ouertani, K. Grebici, and L. Gzara, "A framework to re-organize design activities during engineering change process," in 2007, CDROM. Accessed: Jul. 16 2007. [Online]. Available: https://hal.archives-ouvertes.fr/hal-00163122
68. K. R. Reddi and Y. B. Moon, "System dynamics modeling of engineering change management in a collaborative environment," *Int J Adv Manuf Technol*, vol. 55, 9–12, pp. 1225–1239, 2011, https://doi.org/10.1007/s00170-010-3143-z.
69. K. R. Reddi and Y. B. Moon, "Modelling engineering change management in a new product development supply chain," *International Journal of Production Research*, vol. 51, no. 17, pp. 5271–5291, 2013, https://doi.org/10.1080/00207543.2013.807954.
70. D. Xue, S. Y. Cheing, and P. Gu, "Parameter design considering the impact of design changes on downstream processes based upon the Taguchi method," *Journal of Engineering Design*, vol. 19, no. 4, pp. 299–319, 2008, https://doi.org/10.1080/09544820701376654.
71. S. J. Chinn and G. R. Madey, "Temporal representation and reasoning for workflow in engineering design change review," *IEEE Trans. Eng. Manage.*, vol. 47, no. 4, pp. 485–492, 2000, https://doi.org/10.1109/17.895343.
72. L. F. Hinojos A., N. Iakymenko, and E. Alfnes, "Engineering Changes in the Engineer-to-Order Industry: Challenges of Implementation," in *Lecture Notes in Electrical Engineering, Advanced Manufacturing and Automation IX*, Y. Wang, K. Martinsen, T. Yu, and K. Wang, Eds., Singapore: Springer Singapore, 2020, pp. 662–670.
73. M. Z. Ouertani, "Engineering Change Impact on Product Development Processes," *Syst. Res. Forum*, vol. 03, no. 01, pp. 25–37, 2009, https://doi.org/10.1142/S1793966609000043.
74. S. D. Shivankar, G. M. Kakandikar, and V. M. Nandedkar, "Implementing engineering change management through product life cycle management in automotive field," *IJPLM*, vol. 8, no. 2, p. 132, 2015, https://doi.org/10.1504/ijplm.2015.070579.
75. K. B. Clark and T. Fujimoto, *Product development performance: Strategy, organization, and management in the world auto industry*. Boston, Mass.: Harvard Business School Press, 2005.
76. I. Ullah, D. Tang, and L. Yin, "Engineering Product and Process Design Changes: A Literature Overview," *Procedia CIRP*, vol. 56, pp. 25–33, 2016, https://doi.org/10.1016/j.procir.2016.10.010.
77. F. N. Yeasin, M. Grenn, and B. Roberts, "A Bayesian Networks Approach to Estimate Engineering Change Propagation Risk and Duration," *IEEE Trans. Eng. Manage.*, vol. 67, no. 3, pp. 869–884, 2020, https://doi.org/10.1109/tem.2018.2884242.
78. A. Camarillo, J. Ríos, and K.-D. Althoff, "Agent Based Framework to Support Manufacturing Problem Solving Integrating Product Lifecycle Management and Case-Based Reasoning," in *IFIP Advances in Information and Communication Technology, Product Lifecycle Management and the Industry of the Future*, J. Ríos, A. Bernard, A. Bouras, and S. Foufou, Eds., Cham: Springer International Publishing, 2017, pp. 116–128.
79. P. Burggräf, J. Wagner, and T. M. Saßmannshausen, "Sustainable Interaction of Human and Artificial Intelligence in Cyber Production Management Systems," in *Lecture Notes in Production Engineering, Production at the leading edge of technology*, B.-A. Behrens, A. Brosius, W. Hintze, S. Ihlenfeldt, and J. P. Wulfsberg, Eds., Berlin, Heidelberg: Springer Berlin Heidelberg, 2021, pp. 508–517.

80. R. Baskerville, A. Baiyere, S. Gregor, A. Hevner, and M. Rossi, "Design Science Research Contributions: Finding a Balance between Artifact and Theory," *Journal of the Association for Information Systems*, vol. 19, pp. 358–376, 2018, https://doi.org/10.17705/1jais.00495.
81. J. vom Brocke, R. Winter, A. Hevner, and A. Maedche, "Special Issue Editorial –Accumulation and Evolution of Design Knowledge in Design Science Research: A Journey Through Time and Space," *JAIS*, vol. 21, no. 3, pp. 520–544, 2020, https://doi.org/10.17705/1jais.00611.
82. J. vom Brocke *et al.*, "Reconstructing the Giant: On the Importance of Rigour in Documenting the Literature Search Process," *ECIS 2009 Proceedings 161*, 2009. https://aisel.aisnet.org/ecis2009/161
83. A. Hevner, "A Three Cycle View of Design Science Research," *Scandinavian Journal of Information Systems*, vol. 19, Iss. 2, Article 4, 2007. Available: https://aisel.aisnet.org/sjis/vol19/iss2/4
84. J. Kukulies, B. Falk, and R. H. Schmitt, "Development of Optimized Test Planning Procedures for Stabilizing Ramp-up Processes by Means of Design Science Research," *Procedia CIRP*, vol. 51, pp. 93–98, 2016, https://doi.org/10.1016/j.procir.2016.05.056.
85. R. Haibing, L. Ting, L. Yupeng, and H. Jie, "Multi -source design change propagation path optimisation based on the multi-view complex network model," *Journal of Engineering Design*, vol. 32, no. 1, pp. 28–60, 2021, https://doi.org/10.1080/09544828.2020.1858474.
86. D. Bzdok, N. Altman, and M. Krzywinski, "Statistics versus machine learning," *Nature methods*, vol. 15, no. 4, pp. 233–234, 2018, https://doi.org/10.1038/nmeth.4642.
87. W. Ertel, *Grundkurs Künstliche Intelligenz: Eine praxisorientierte Einführung*, 4th ed. Wiesbaden: Springer Vieweg, 2016.
88. P. Kumar and Ratneshwer, "Some Observations on Dependency Analysis of SOA Based Systems," *IJITCS*, vol. 8, no. 1, pp. 54–66, 2016, https://doi.org/10.5815/ijitcs.2016.01.07.
89. S. Helms, F. Behncke, and Lindl, "Classification of Methods for the Indication of Change Propagation - A Literature Review," *DS 77: Proceedings of the DESIGN 2014 13th International Design Conference*, pp. 211–220, 2014.
90. I. Ullah, D. Tang, and L. Yin, "Engineering Change Implications on Product Design: A Review of the Literature," in *Proceedings of the 2015 International Conference on Education, Management and Computing Technology*, Tianjin, China, 2015.
91. M. Masmoudi, P. Leclaire, M. Zolghadri, and M. Haddar, "Engineering Change Management (ECM) Methods: Classification According to Their Dependency Models," in *Lecture Notes in Mechanical Engineering, Design and Modeling of Mechanical Systems—III*, M. Haddar, F. Chaari, A. Benamara, M. Chouchane, C. Karra, and N. Aifaoui, Eds., Cham: Springer International Publishing, 2018, pp. 1169–1178.
92. A. Tale-Yazdi, N. Kattner, L. Becerril, and U. Lindemann, "A Literature Review on Approaches for the Retrospective Utilisation of Data in Engineering Change Management," in *2018 IEEE International Conference 2018*, 2018, pp. 612–616.
93. A. S. Balakrishnan and J. Suresh, "Engineering changes - research findings and future directions," *IJENM*, vol. 13, no. 1, p. 66, 2022, https://doi.org/10.1504/IJENM.2022.122418.
94. E. F. Colombo, G. Cascini, and O. L. de Weck, "Classification of Change-Related Ilities Based on a Literature Review of Engineering Changes," *JID*, vol. 20, no. 4, pp. 3–23, 2017, https://doi.org/10.3233/jid-2016-0019.

95. R. B. Singh, A. S. Baghel, and A. Agarwal, "A review on VLSI floorplanning optimization using metaheuristic algorithms," in *2016 International Conference on Electrical, Electronics, and Optimization Techniques (ICEEOT)*, Chennai, India, 2016, pp. 4198–4202.
96. H. M. Cooper, "Organizing knowledge syntheses: A taxonomy of literature reviews," (in En;en), *Knowledge in Society*, vol. 1, no. 1, pp. 104–126, 1988, https://doi.org/10.1007/BF03177550.
97. J. Rowley and F. Slack, "Conducting a literature review," *Management Research News*, vol. 27, no. 6, pp. 31–39, 2004, https://doi.org/10.1108/01409170410784185.
98. M. Gusenbauer and N. R. Haddaway, "Which academic search systems are suitable for systematic reviews or meta-analyses? Evaluating retrieval qualities of Google Scholar, PubMed, and 26 other resources," *Research synthesis methods*, vol. 11, no. 2, pp. 181–217, 2020, https://doi.org/10.1002/jrsm.1378.
99. D. Moher, A. Liberati, J. Tetzlaff, and D. G. Altman, "Preferred reporting items for systematic reviews and meta-analyses: the PRISMA statement," *International journal of surgery (London, England)*, vol. 8, no. 5, pp. 336–341, 2010, https://doi.org/10.1016/j.ijsu.2010.02.007.
100. J. Webster and R. T. Watson, "Analyzing the Past to Prepare for the Future: Writing a Literature Review," *MIS Quarterly*, vol. 26, no. 2, pp. xiii–xxiii, 2002. [Online]. Available: http://www.jstor.org/stable/4132319
101. K. Pasupa and W. Sunhem, "A comparison between shallow and deep architecture classifiers on small dataset," in *2016 8th International Conference on Information Technology and Electrical Engineering (ICITEE)*, Yogyakarta, Indonesia, 2016, pp. 1–6.
102. M. Feurer and F. Hutter, "Hyperparameter Optimization," in *The Springer Series on Challenges in Machine Learning, Automated Machine Learning*, F. Hutter, L. Kotthoff, and J. Vanschoren, Eds., Cham: Springer International Publishing, 2019, pp. 3–33.
103. R. Wirth and J. Hipp, "Crisp-dm: towards a standard process modell for data mining," in *Proceedings of the Fourth International Conference on the Practical Application of Knowledge Discovery and Data Mining*, 2000.
104. M. Riesener, C. Dolle, M. Mendl-Heinisch, and G. Schuh, "Applying the Random Forest Algorithm to Predict Engineering Change Effort," in *2021 IEEE Technology & Engineering Management Conference - Europe (TEMSCON-EUR)*, Dubrovnik, Croatia, 2021, pp. 1–6.
105. M. Pacella, A. Grieco, and M. Blaco, "On the Use of Self-Organizing Map for Text Clustering in Engineering Change Process Analysis: A Case Study," *Computational intelligence and neuroscience*, vol. 2016, p. 5139574, 2016, https://doi.org/10.1155/2016/5139574.
106. C. Mehta, L. Patil, and D. Dutta, "An Approach to Determine Important Attributes for Engineering Change Evaluation," *Journal of Mechanical Design*, vol. 135, no. 4, 2013a, https://doi.org/10.1115/1.4023551.
107. J. Chen, S. Zhang, M. Wang, and C. Xu, "A novel change feature-based approach to predict the impact of current proposed engineering change," *Advanced Engineering Informatics*, vol. 33, pp. 132–143, 2017, https://doi.org/10.1016/j.aei.2017.06.002.
108. B. Morkos, J. Mathieson, and J. D. Summers, "Comparative analysis of requirements change prediction models: manual, linguistic, and neural network," *Research in Engineering Design*, vol. 25, no. 2, pp. 139–156, 2014, https://doi.org/10.1007/s00163-014-0170-z.

109. Y. Pan and R. Stark, "An Ensemble Learning based Hierarchical Multi-label Classification Approach to Identify Impacts of Engineering Changes," in *2020 IEEE 32nd International Conference on Tools with Artificial Intelligence (ICTAI)*, Baltimore, MD, USA, 2020, pp. 1260–1267.
110. A. Sharafi, F. Elezi, F. Zuber, P. Wolf, H. Krcmar, and U. Lindemann, "Determining the Drivers for Long Lead Times of Engineering Change Orders: A Data Mining Approach," *DS 70: Proceedings of DESIGN 2012, the 12th International Design Conference, Dubrovnik, Croatia*, pp. 299–310, 2012.
111. J. Heaton, S. McElwee, J. Fraley, and J. Cannady, "Early stabilizing feature importance for TensorFlow deep neural networks," in *2017 International Joint Conference on Neural Networks (IJCNN)*, Anchorage, AK, USA, 2017, pp. 4618–4624.
112. A. Grieco, M. Pacella, and M. Blaco, "On the Application of Text Clustering in Engineering Change Process," *Procedia CIRP*, vol. 62, pp. 187–192, 2017, https://doi.org/10.1016/j.procir.2016.06.019.
113. R.-S. Lu, Z.-T. Wu, K.-W. Peng, and T.-Y. Yu, "Use of the self-organizing feature map to diagnose abnormal engineering change," in *Seventh International Conference on Digital Image Processing (ICDIP 2015)*, Los Angeles, United States, 2015, p. 963119.
114. Z. Li, X. Sun, X. Chen, Y. Zhang, Q. Li, and C. Peng, "Model Construction for Complex Product Design Change with Improved Dendritic Neural Network," in *2021 IEEE International Conference on Recent Advances in Systems Science and Engineering (RASSE)*, Shanghai, China, 2021b, pp. 1–8.
115. Y. Todo, H. Tamura, K. Yamashita, and Z. Tang, "Unsupervised learnable neuron model with nonlinear interaction on dendrites," *Neural networks : the official journal of the International Neural Network Society*, vol. 60, pp. 96–103, 2014, https://doi.org/10.1016/j.neunet.2014.07.011.
116. I. Ö. Arnarsson, O. Frost, E. Gustavsson, D. Stenholm, M. Jirstrand, and J. Malmqvist, "Supporting Knowledge Re-Use with Effective Searches of Related Engineering Documents—A Comparison of Search Engine and Natural Language Processing-Based Algorithms," *Proc. Int. Conf. Eng. Des.*, vol. 1, no. 1, pp. 2597–2606, 2019, https://doi.org/10.1017/dsi.2019.266.
117. I. Ö. Arnarsson, O. Frost, E. Gustavsson, M. Jirstrand, and J. Malmqvist, "Natural language processing methods for knowledge management—Applying document clustering for fast search and grouping of engineering documents," *Concurrent Engineering*, vol. 29, no. 2, pp. 142–152, 2021, https://doi.org/10.1177/1063293X20982973.
118. M. Riesener, C. Dölle, M. Mendl-Heinisch, G. Schuh, and A. Keuper, "Derivation of Description Features for Engineering Change Request by Aid of Latent Dirichlet Allocation," *Proc. Des. Soc.: Des. Conf.*, vol. 1, pp. 697–706, 2020, https://doi.org/10.1017/dsd.2020.98.
119. E. C. Koh, "Design change prediction based on social media sentiment analysis," *AIEDAM*, vol. 36, 2022, https://doi.org/10.1017/S0890060422000129.
120. M. Fathianathan, A. S. Kumar, and A. Y. C. Nee, "An Adaptive Machining Fixture Design System for Automatically Dealing With Design Changes," *Journal of Computing and Information Science in Engineering*, vol. 7, no. 3, pp. 259–268, 2007, https://doi.org/10.1115/1.2752816.
121. Y. Li, W. Zhao, and J. Zhang, "Resource-constrained scheduling of design changes based on simulation of change propagation process in the complex engineering design,"

Research in Engineering Design, vol. 30, no. 1, pp. 21–40, 2019c, https://doi.org/10.1007/s00163-018-0302-y.

122. Y. Li and W. Zhao, "An integrated change propagation scheduling approach for product design," *Concurrent Engineering*, vol. 22, no. 4, pp. 347–360, 2014, https://doi.org/10.1177/1063293X14553809.
123. X. Lian, Y. Yang, and J. Wang, "Research on complex product design change propagation based on complex networks," in *2017 6th International Conference on Industrial Technology and Management (ICITM)*, Cambridge, United Kingdom, 2017, pp. 80–84.
124. S.-T. Wang, "Integration of a GA and PSO for discussing the impact of 3C product engineering changes on customisation degree," *International Journal of Production Research*, vol. 50, no. 15, pp. 4224–4236, 2012, https://doi.org/10.1080/00207543.2011.603708.
125. H.-S. Wang and Z.-H. Che, "Applying and Comparing Four Different PSO Approaches in Integrated Problem of Product Change Planning, Part Supplier Selection, and Quantity Allocation," *Journal of the Chinese Institute of Industrial Engineers*, vol. 26, no. 2, pp. 87–98, 2009, https://doi.org/10.1080/10170660909509125.
126. C. Mehta, L. Patil, and D. Dutta, "An Information-Based Approach to Compute Similarity Between Engineering Changes," *IEEE Trans. Automat. Sci. Eng.*, vol. 9, no. 2, pp. 330–341, 2012b, https://doi.org/10.1109/TASE.2011.2176538.
127. C. Mehta, L. Patil, and D. Dutta, "An Approach to Predict Impact of Proposed Engineering Change Effect," *Journal of Computing and Information Science in Engineering*, vol. 12, no. 2, 2012c, https://doi.org/10.1115/1.4005593.
128. J. Lee and Y. S. Hong, "Bayesian network approach to change propagation analysis," *Research in Engineering Design*, vol. 28, no. 4, pp. 437–455, 2017, https://doi.org/10.1007/s00163-017-0252-9.
129. S. Mirdamadi, S.-A. Addouche, and M. Zolghadri, "A Bayesian approach to model change propagation mechanisms," *Procedia CIRP*, vol. 70, pp. 1–6, 2018, https://doi.org/10.1016/j.procir.2018.03.309.
130. T. M. L. Diallo and M. Zolghadri, "A Causal Dependencies Identification and Modelling Approach for Redesign Process," in *IFIP Advances in Information and Communication Technology, Product Lifecycle Management to Support Industry 4.0*, P. Chiabert, A. Bouras, F. Noël, and J. Ríos, Eds., Cham: Springer International Publishing, 2018, pp. 778–788.
131. Y. Damak, Y. Leroy, G. Trehard, and M. Jankovic, "Operational Context Change Propagation Prediction on Autonomous Vehicles Architectures," *Journal of Autonomous Vehicles and Systems*, vol. 1, no. 3, 2021, https://doi.org/10.1115/1.4052556.
132. J. Hu and M.-A. Cardin, "Generating flexibility in the design of engineering systems to enable better sustainability and lifecycle performance," *Research in Engineering Design*, vol. 26, no. 2, pp. 121–143, 2015, https://doi.org/10.1007/s00163-015-0189-9.
133. C. Aguwa, D. Egeonu, E.-E. Etu, and L. Monplaisir, "Fuzzy-Based Integrated Customer Satisfaction Index to Enable Engineering Change," in *Proceedings of the 2017 Industrial and Systems Engineering Conference*, 2017, pp. 1036–1041.
134. K. Sandkuhl, A. Smirnov, and N. Shilov, "Information Logistics in Engineering Change Management: Integrating Demand Patterns and Recommendation Systems," in *Lecture Notes in Business Information Processing, Workshops on Business Informatics*

Research, W. van der Aalst et al., Eds., Berlin, Heidelberg: Springer Berlin Heidelberg, 2012, pp. 14–25.

135. T. M. Kindsmuller *et al.*, "Mitigating the effort for engineering changes in product development using a fuzzy expert system," in *2014 IEEE International Conference on Industrial Engineering and Engineering Management*, Selangor Darul Ehsan, Malaysia, 2014, pp. 602–606.
136. M. Wooldridge, *An introduction to multiagent systems,* 2nd ed. [?]: Wiley-Blackwell, 2003.
137. J. Bender, S. Kehl, and J. P. Müller, "A Comparison of Agent-Based Coordination Architecture Variants for Automotive Product Change Management," in *Lecture Notes in Computer Science, Multiagent System Technologies*, J. P. Müller, W. Ketter, G. Kaminka, G. Wagner, and N. Bulling, Eds., Cham: Springer International Publishing, 2015, pp. 249–267.
138. B. Beroule, A.-J. Fougeres, and E. Ostrosi, "Engineering change management through consensus seeking by fuzzy agents," in *Second World Conference on Complex Systems (WCCS)*, Agadir, Morocco, 2014, pp. 542–547.
139. S. Ma, Z. Jiang, and W. Liu, "Multi-variation propagation prediction based on multi-agent system for complex mechanical product design," *Concurrent Engineering*, vol. 25, no. 4, pp. 316–330, 2017, https://doi.org/10.1177/1063293X17708820.
140. D. Habhouba, A. Desrochers, and S. Cherkaoui, "Agent-based assistance for engineering change management: An implementation prototype," in *2009 13th International Conference on Computer Supported Cooperative Work in Design*, Santiago, Chile, 2009, pp. 288–293.
141. S. Oh, Y. Jung, S. Kim, I. Lee, and N. Kang, "Deep Generative Design: Integration of Topology Optimization and Generative Models," *Journal of Mechanical Design*, vol. 141, no. 11, 2019, https://doi.org/10.1115/1.4044229.
142. Y. Yu, T. Hur, J. Jung, and I. G. Jang, "Deep learning for determining a near-optimal topological design without any iteration," *Struct Multidisc Optim*, vol. 59, no. 3, pp. 787–799, 2019, https://doi.org/10.1007/s00158-018-2101-5.
143. T. Fässberg *et al.*, "An Empirical Study Towards a Definition of Production Complexity," in *Proceedings of the 21st International Conference on Production Research (ICPR)*, 2011.
144. E.-G. Talbi, *Metaheuristics: From design to implementation*. Hoboken, NJ: John Wiley & Sons, 2009. [Online]. Available: https://ebookcentral.proquest.com/lib/kxp/detail.action?docID=448946
145. C. Blum and A. Roli, "Metaheuristics in combinatorial optimization," *ACM Computing Surveys*, vol. 35, no. 3, pp. 268–308, 2003, https://doi.org/10.1145/937503.937505.
146. A. Ahmid, T.-M. Dao, and van LÊ, "Comparison Study of Discrete Optimization Problem Using Meta-Heuristic Approaches: A Case Study," *International Journal of Industrial Engineering and Operations Management*, vol. 01, no. 02, 2019, https://doi.org/10.46254/j.ieom.20190201.
147. A. Alfageme, I. Esnaola-Gonzalez, F. J. Díez, and E. Gilabert, "Metaheuristics for Optimal Scheduling of Appliances in Energy Efficient Neighbourhoods," in *Lecture Notes in Computer Science, Progress in Artificial Intelligence*, G. Marreiros, F. S. Melo, N. Lau, H. Lopes Cardoso, and L. P. Reis, Eds., Cham: Springer International Publishing, 2021, pp. 151–162.

148. F.-S. Hsieh, "A Comparative Study of Several Metaheuristic Algorithms to Optimize Monetary Incentive in Ridesharing Systems," *ISPRS International Journal of Geo-Information*, vol. 9, no. 10, p. 590, 2020, https://doi.org/10.3390/ijgi9100590.
149. V. Beiranvand, W. Hare, and Y. Lucet, "Best practices for comparing optimization algorithms," *Optimization and Engineering*, vol. 18, no. 4, pp. 815–848, 2017, https://doi.org/10.1007/s11081-017-9366-1.
150. Y. L. Yin and T. Tang, "Research on Manufacturing Engineering Change Implementation Planning," *Materials and Manufacturing Research*, vol. 658, pp. 460–463, 2013, https://doi.org/10.4028/www.scientific.net/AMR.658.460.
151. H. Stadler and C. Kilger, *Supply chain management and advanced planning: Concepts, models, software, and case studies,* 4th ed. Berlin: Springer, 2010.
152. O. W. Wight, *Manufacturing resource planning: MRP II ; unlocking America's productivity potential*. Essex Junction, VT: Wight, 1984.
153. F. Glover, "Tabu Search—Part I," *ORSA Journal on Computing*, vol. 1, no. 3, pp. 190–206, 1989, https://doi.org/10.1287/ijoc.1.3.190.
154. S. Kirkpatrick, C. D. Gelatt, and M. P. Vecchi, "Optimization by simulated annealing," *Science (New York, N.Y.)*, vol. 220, no. 4598, pp. 671–680, 1983, https://doi.org/10.1126/science.220.4598.671.
155. J. H. Holland, "Outline for a Logical Theory of Adaptive Systems," *J. ACM*, vol. 9, no. 3, pp. 297–314, 1962, https://doi.org/10.1145/321127.321128.
156. J. H. Holland, *Adaptation in Natural and Artificial Systems*: The MIT Press, 1992.
157. M. Dorigo, V. Maniezzo, and A. Colorni, "Ant system: optimization by a colony of cooperating agents," *IEEE transactions on systems, man, and cybernetics. Part B, Cybernetics : a publication of the IEEE Systems, Man, and Cybernetics Society*, vol. 26, no. 1, pp. 29–41, 1996, https://doi.org/10.1109/3477.484436.
158. J. Kennedy and R. Eberhart, "Particle swarm optimization," in *Proceedings of ICNN'95 - International Conference on Neural Networks*, Perth, WA, Australia, 1995, pp. 1942–1948.
159. A. C. B. Monteiro, R. P. França, R. Arthur, and Y. Iano, "The fundamentals and potential of heuristics and metaheuristics for multiobjective combinatorial optimization problems and solution methods," in *Multi-Objective Combinatorial Optimization Problems and Solution Methods*, M. Toloo, Ed., San Diego: Elsevier Science & Technology, 2022, pp. 9–29.
160. M. Kalami Heris, *Practical Genetic Algorithms in Python and MATLAB – Video Tutorial.* [Online]. Available: https://yarpiz.com/632/ypga191215-practical-genetic-algorithms-in-python-and-matlab
161. J. Demšar, "Statistical Comparisons of Classifiers over Multiple Data Sets," *J. Mach. Learn. Res.*, vol. 7, pp. 1–30, 2006, https://doi.org/10.5555/1248547.1248548.
162. M. Friedman, "The Use of Ranks to Avoid the Assumption of Normality Implicit in the Analysis of Variance," *Journal of the American Statistical Association*, vol. 32, no. 200, pp. 675–701, 1937, https://doi.org/10.1080/01621459.1937.10503522.
163. P. B. Nemenyi, *Distribution-free multiple comparisons*, 1963.
164. P. Burggräf, F. Steinberg, T. Weißer, and O. Radisic-Aberger, "Deciding on when to change – a benchmark of metaheuristic algorithms for timing engineering changes," *International Journal of Production Research*, pp. 1–21, 2023b, https://doi.org/10.1080/00207543.2023.2226778.

165. F. R. do Nascimento, A. G. dos Santos, L. A. Da Fonseca Junior, and D. M. Nunes, "Application of Lean Six Sigma to Reduce Delays in Engineering Changes," *IEEE Trans. Eng. Manage.*, pp. 1–14, 2022, https://doi.org/10.1109/TEM.2022.3199388.
166. N. Boff Medeiros, F. S. Fogliatto, M. Karla Rocha, and G. L. Tortorella, "Predicting the length-of-stay of pediatric patients using machine learning algorithms," *International Journal of Production Research*, pp. 1–14, 2023, https://doi.org/10.1080/00207543.2023.2235029.
167. N. Tax, I. Verenich, M. La Rosa, and M. Dumas, "Predictive Business Process Monitoring with LSTM Neural Networks," in *Lecture Notes in Computer Science, Advanced Information Systems Engineering*, E. Dubois and K. Pohl, Eds., Cham: Springer International Publishing, 2017, pp. 477–492.
168. P. Burggraf, J. Wagner, B. Heinbach, and F. Steinberg, "Machine Learning-Based Prediction of Missing Components for Assembly – a Case Study at an Engineer-to-Order Manufacturer," *IEEE Access*, vol. 9, pp. 105926–105938, 2021, https://doi.org/10.1109/ACCESS.2021.3075620.
169. F. M. Maggi, C. Di Francescomarino, M. Dumas, and C. Ghidini, "Predictive Monitoring of Business Processes," in 2014, pp. 457–472. [Online]. Available: https://link.springer.com/chapter/10.1007/978-3-319-07881-6_31
170. I. Teinemaa, M. Dumas, A. Leontjeva, and F. M. Maggi, "Temporal stability in predictive process monitoring," *Data Min Knowl Disc*, vol. 32, no. 5, pp. 1306–1338, 2018a, https://doi.org/10.1007/s10618-018-0575-9.
171. W. Kratsch, J. Manderscheid, M. Röglinger, and J. Seyfried, "Machine Learning in Business Process Monitoring: A Comparison of Deep Learning and Classical Approaches Used for Outcome Prediction," *Bus Inf Syst Eng*, vol. 63, no. 3, pp. 261–276, 2021, https://doi.org/10.1007/s12599-020-00645-0.
172. K. Peffers, T. Tuunanen, M. A. Rothenberger, and S. Chatterjee, "A Design Science Research Methodology for Information Systems Research," *Journal of Management Information Systems*, vol. 24, no. 3, pp. 45–77, 2007, https://doi.org/10.2753/MIS0742-1222240302.
173. S. Gregor and A. R. Hevner, "Positioning and Presenting Design Science Research for Maximum Impact," *MISQ*, vol. 37, no. 2, pp. 337–355, 2013, https://doi.org/10.25300/MISQ/2013/37.2.01.
174. L. Breiman, "Random Forests," (in En;en), *Machine Learning*, vol. 45, no. 1, pp. 5–32, 2001, https://doi.org/10.1023/A:1010933404324.
175. T. Chen and C. Guestrin, "XGBoost," in *Proceedings of the 22nd ACM SIGKDD International Conference on Knowledge Discovery and Data Mining*, San Francisco California USA, 2016, pp. 785–794.
176. S. Hochreiter and J. Schmidhuber, "Long short-term memory," *Neural computation*, vol. 9, no. 8, pp. 1735–1780, 1997, https://doi.org/10.1162/neco.1997.9.8.1735.
177. J. Evermann, J.-R. Rehse, and P. Fettke, "A Deep Learning Approach for Predicting Process Behaviour at Runtime," in *Lecture Notes in Business Information Processing, Business Process Management Workshops*, M. Dumas and M. Fantinato, Eds., Cham: Springer International Publishing, 2017, pp. 327–338.
178. D. C. Wynn, N. H. M. Caldwell, and P. John Clarkson, "Predicting Change Propagation in Complex Design Workflows," *Journal of Mechanical Design*, vol. 136, no. 8, 2014b, https://doi.org/10.1115/1.4027495.

179. I. Teinemaa, N. Tax, M. de Leoni, M. Dumas, and F. M. Maggi, "Alarm-Based Prescriptive Process Monitoring," in *Lecture Notes in Business Information Processing, Business Process Management Forum*, M. Weske, M. Montali, I. Weber, and J. vom Brocke, Eds., Cham: Springer International Publishing, 2018b, pp. 91–107.
180. C. Drummond and R. C. Holte, "Cost curves: An improved method for visualizing classifier performance," *Mach Learn*, vol. 65, no. 1, pp. 95–130, 2006, https://doi.org/10.1007/s10994-006-8199-5.
181. M. de Leoni, W. M. van der Aalst, and M. Dees, "A general process mining framework for correlating, predicting and clustering dynamic behavior based on event logs," *Information Systems*, vol. 56, pp. 235–257, 2016, https://doi.org/10.1016/j.is.2015.07.003.
182. H. Weytjens and J. de Weerdt, "Process Outcome Prediction: CNN vs. LSTM (with Attention)," in 2020, pp. 321–333. [Online]. Available: https://link.springer.com/chapter/10.1007/978-3-030-66498-5_24
183. D. Lichtenwalter, P. Burggräf, J. Wagner, and T. Weißer, "Deep multimodal learning for manufacturing problem solving," *Procedia CIRP*, vol. 99, pp. 615–620, 2021, https://doi.org/10.1016/j.procir.2021.03.083.
184. J. Bergstra and Y. Bengio, "Random Search for Hyper-Parameter Optimization," *Journal of Machine Learning Research*, vol. 13, no. 10, pp. 281–305, 2012.
185. A. P. Bradley, "The use of the area under the ROC curve in the evaluation of machine learning algorithms," *Pattern Recognition*, vol. 30, no. 7, pp. 1145–1159, 1997, https://doi.org/10.1016/S0031-3203(96)00142-2.
186. J. Davis and M. Goadrich, "The relationship between Precision-Recall and ROC curves," in *Proceedings of the 23rd international conference on Machine learning*, Pittsburgh, Pennsylvania, 2006, pp. 233–240.
187. N. M. Adams and D. J. Hand, "Comparing classifiers when the misallocation costs are uncertain," *Pattern Recognition*, vol. 32, no. 7, pp. 1139–1147, 1999, https://doi.org/10.1016/S0031-3203(98)00154-X.
188. T. Fawcett, "ROC Graphs: Notes and Practical Considerations for Data Mining Researchers," *ReCALL*, vol. 31, pp. 1–38, 2004.
189. J. Hernández-Orallo, P. Flach, and C. Ferri, "A Unified View of Performance Metrics: Translating Threshold Choice into Expected Classification Loss," *J. Mach. Learn. Res.*, vol. 13, no. 1, pp. 2813–2869, 2012.
190. D. McElfresh *et al.*, "When Do Neural Nets Outperform Boosted Trees on Tabular Data?," in *Advances in Neural Information Processing Systems*, 2023, pp. 76336–76369.
191. L. Grinsztajn, E. Oyallon, and G. Varoquaux, "Why do tree-based models still outperform deep learning on typical tabular data?," in *Advances in Neural Information Processing Systems*, 2022, pp. 507–520.
192. P. K. Sriram, H. C. Dreyer, and E. Alfnes, "Understanding Key Engineering Changes for Materials Management in ETO Environment," in *IFIP Advances in Information and Communication Technology, Advances in Production Management Systems: Innovative Production Management Towards Sustainable Growth*, S. Umeda, M. Nakano, H. Mizuyama, H. Hibino, D. Kiritsis, and G. von Cieminski, Eds., Cham: Springer International Publishing, 2015, pp. 256–262.

193. P. Jonsson, J. Öhlin, H. Shurrab, J. Bystedt, A. Sheikh Muhammad, and V. Verendel, "What are the root causes of material delivery schedule inaccuracy in supply chains?," *IJOPM*, vol. 44, no. 13, pp. 34–68, 2024, https://doi.org/10.1108/IJOPM-12-2022-0806.
194. F. Steinberg, P. Burggräf, J. Wagner, B. Heinbach, T. Saßmannshausen, and A. Brintrup, "A novel machine learning model for predicting late supplier deliveries of low-volume-high-variety products with application in a German machinery industry," *Supply Chain Analytics*, vol. 1, p. 100003, 2023, https://doi.org/10.1016/j.sca.2023.100003.
195. G. T. Lakshmanan, S. Duan, P. T. Keyser, F. Curbera, and R. Khalaf, "Predictive Analytics for Semi-structured Case Oriented Business Processes," in *Lecture Notes in Business Information Processing, Business Process Management Workshops*, M. zur Muehlen and J. Su, Eds., Berlin, Heidelberg: Springer Berlin Heidelberg, 2011, pp. 640–651.
196. N. Mehdiyev, M. Majlatow, and P. Fettke, "Quantifying and explaining machine learning uncertainty in predictive process monitoring: an operations research perspective," *Ann Oper Res*, 2024, https://doi.org/10.1007/s10479-024-05943-4.
197. J. Wang, V. Chang, D. Yu, C. Liu, X. Ma, and D. Yu, "Conformance-oriented Predictive Process Monitoring in BPaaS Based on Combination of Neural Networks," *J Grid Computing*, vol. 20, no. 3, 2022, https://doi.org/10.1007/s10723-022-09613-2.
198. Z. Xing, J. Pei, G. Dong, and P. S. Yu, "Mining Sequence Classifiers for Early Prediction," in *Proceedings of the 2008 SIAM International Conference on Data Mining*, 2008, pp. 644–655.
199. A. Leontjeva, R. Conforti, C. Di Francescomarino, M. Dumas, and F. M. Maggi, "Complex Symbolic Sequence Encodings for Predictive Monitoring of Business Processes," in *Lecture Notes in Computer Science, Business Process Management*, H. R. Motahari-Nezhad, J. Recker, and M. Weidlich, Eds., Cham: Springer International Publishing, 2015, pp. 297–313.

MIX
Papier aus verantwortungsvollen Quellen
Paper from responsible sources
FSC® C105338

If you have any concerns about our products,
you can contact us on
ProductSafety@springernature.com

In case Publisher is established outside the EU,
the EU authorized representative is:
Springer Nature Customer Service Center GmbH
Europaplatz 3, 69115 Heidelberg, Germany

Printed by Libri Plureos GmbH
in Hamburg, Germany